Introduction
to Engineering Technology

Introduction to Engineering Technology

FOURTH EDITION

Robert J. Pond
Central Ohio Technical College

Prentice Hall
Upper Saddle River, New Jersey   *Columbus, Ohio*

Library of Congress Cataloging-in-Publication Data

Pond, Robert J.
 Introduction to engineering technology / Robert J. Pond. — 4th
ed.
   p. cm.
 Includes bibliographical references and index.
 ISBN 0-13-854812-9 (pbk.)
 1. Engineering—Vocational guidance. I. Title.
TA157.P66 1999
620'.0023—dc21   98-15343
                    CIP

Cover Photo: Stock Market
Editor: Stephen Helba
Production Editor: Patricia S. Kelly
Design Coordinator: Karrie M. Converse
Text Designer: Angela Foote
Cover Designer: Raymond Hummons
Production Manager: Deidra M. Schwartz
Marketing Manager: Frank Mortimer

This book was set in ITC Garamond and Helvetica by Carlisle Communications Ltd. and was printed and bound by R.R. Donnelley & Sons Company. The cover was printed by Phoenix Color Corp.

 © 1999, 1996 by Prentice-Hall, Inc.
Simon & Schuster/A Viacom Company
Upper Saddle River, New Jersey 07458

Earlier editions © 1993 by Macmillan Publishing Company and © 1990 by Merrill Publishing Company.

Printed in the United States of America

10 9 8 7 6 5 4 3 2 1

ISBN: 0-13-854812-9

Prentice-Hall International (UK) Limited, *London*
Prentice-Hall of Australia Pty. Limited, *Sydney*
Prentice-Hall of Canada, Inc., *Toronto*
Prentice-Hall Hispanoamericana, S.A., *Mexico*
Prentice-Hall of India Private Limited, *New Delhi*
Prentice-Hall of Japan, Inc., *Tokyo*
Simon & Schuster Asia Pte. Ltd., *Singapore*
Editora Prentice-Hall do Brasil, Ltda., *Rio de Janeiro*

*Dedicated to the technologists of the 21st century
and their contributions to a safe, abundant,
and harmonious society.*

Preface

Graduates of 2- and 4-year technical programs will become society's *implementors,* those who transform concepts into reality. They must combine a practical understanding of materials, machinery, and processes with the theories of today's practicing engineer. Graduates will work on the cutting edge of technology, installing, operating, and maintaining new equipment in manufacturing and service industries. The implementors will be needed by society, challenged by their careers, and rewarded by their successes. This book is for them.

Changes in the Fourth Edition

The most extensive improvements in the fourth edition of *Introduction to Engineering Technology* are

- Employment and salary information, as well as occupational information, has been updated.
- Cooperative education, preparation for the interview, and the importance of the placement office has been amplified. The importance of early planning for graduation and the *graduation portfolio* are discussed.
- A discussion of the *scientific method* has been added to Chapter 7.
- Technological advances since the third edition have been added. Graphing calculators (e.g., the Tl-85) have been updated in Chapter 4. Computer advances, such as Windows 95, are included in Chapter 8.
- The *Internet Guide* is a new, separate section in Chapter 8.
- A QBASIC example has been added to Chapter 9.
- New developments, such as those in the telecommunications industry, have been added to Chapter 10.
- New figures and photos have been added to clarify the concepts presented.
- Both students and instructors will benefit from the increased number of applications and homework problems.

To the Student

Today, the key to survival—for all nations—is to have an adequate number of technicians and technologists to produce, install, and maintain state-of-the-art equipment, train and supervise industry's skilled workers, and support research and development efforts. This book's holistic approach will help those of you seeking to become society's implementors receive a quality education. It will allow you to answer these questions:

- Am I suited for a career in technology?
- What tools do I need to be successful?
- How will I apply these tools in business and industry?

This text will present an overall picture of the engineering world with the technician's and technologist's places in it. It stresses the importance of possessing a good attitude and paying close attention to detail. It also provides you with the opportunity to use the language and tools of the math-sciences. The language and tools this book presents are those you will need to be successful in the real world of business and industry. I encourage you to sharpen your problem-solving abilities by working the problems at the ends of the chapters. The answers to the odd-numbered problems are at the back of this text.

To the Instructor

The text material has been used in the classroom and has worked well for my two-year college students. Students without a strong mathematics and science background *can* learn from the material: Appendices B and C review the mathematical principles necessary to understand the text. *Introduction to Engineering Technology* may be used as the primary text for Orientation to Engineering Technology courses or as a supplement for courses requiring the use of applied mathematics, computers, or scientific calculators.

The text also may be used in secondary school technical preparatory (tech prep) programs. Dale Parnell's popular book, *The Neglected Majority,* discusses the need for more structured mathematics and science education for the "middle fifty percent of any high school's student body." This text provides a vital practical base to support such a structured approach.

This book contains numerous practical applications to enhance understanding of the concepts discussed. Students wish to work with current applications (e.g., graphing calculators in Chapter 4 and ladder diagrams in Chapter 9) as well as learn the basics. I encourage you to discuss the applications in class, as well as to discuss your own problem-solving experiences.

Introduction to Engineering Technology, Fourth Edition, is supported by a complete instructor's manual, containing performance objectives for each chapter, suggested class activities, transparency masters, and worked-out solutions to all problems.

Organization

Chapters 1 and 2 contain a brief history of engineering and career information for technicians and technologists. Career information includes such topics as the role of the technologist, the need for good communication skills and teamwork, and potential salary information for the future. Major technologies discussed are Chemical, Civil, Architectural, Electrical/Electronic, Computer, Industrial, and Mechanical. Chapter 3 covers college survival skills—using the technical library, maintaining good grades, scheduling adequate study time, and applying basic problem-solving skills. Résumé writing, interviewing techniques, and looking ahead to graduation prepare students for their ultimate goal—gaining desirable employment. Membership in a professional society is recommended. Appendix A provides a listing of those professional societies most responsive to technicians.

Chapter 4 familiarizes students with the calculator, including the use of algebraic logic systems and the mathematics of signed numbers. The rules for adding, subtracting, multiplying, and dividing signed numbers appear in Appendix B. Chapter 5 discusses the use of dimensions and units. A simple four-step approach ensures the student's good grasp of unit conversion. The table of equivalents on the inside front cover provides necessary conversion factors. Chapter 6, which gives the student a taste of the geometry needed in the technologies, is filled with examples that feature step-by-step solutions. The material covering right-triangle trigonometry uses only the first quadrant (acute angles less than 90°). The vector material may be effectively introduced in the first term of study. Appendix C presents mnemonics to aid students in using trigonometric functions.

Chapter 7 focuses on communication, including proper experimental methods, graphing, oral reporting, and report writing. Chapter 8 covers the basics of microcomputers and personal computers. Early exposure to the computer is essential to both today's college success and tomorrow's career satisfaction for the technical student. Two glossaries in the chapter will help the student master the language of computers, and the inside back cover lists the most frequently used (command-oriented) DOS commands and selected operating system commands. The chapter ends with the section "Purchasing Your PC," which addresses the specifications needed to purchase a suitable home computer.

Chapter 9 addresses industrial automation: programming languages such as BASIC are described and computer-integrated manufacturing (CIM), including controllers and control loops, numerical control (NC), flexible manufacturing systems (FMS), and distributed control systems (DCS), is discussed. Appendices D and E will supplement this material. Chapter 10 describes future challenges the engineering technologist will confront: robotics, expert systems, optical systems, new composite materials, and protection of our environment. Appendix F, a glossary of abbreviations and acronyms used in technology, will be a valuable reference throughout the course and beyond.

Acknowledgments

I could not have written this text without the support of my wife Constance. To have an honest and sensitive critic, who is always there, is paramount to any writer's success.

My colleagues at Central Ohio Technical College have been of immeasurable value to all of my professional work. I wish to thank them for their proofreading, comments, and recommendations. Special thanks goes to Darla Sunnenberg for her excellent description of the Windows 95 operating environment.

Special thanks also to Philip Regalbuto, an engineering technology instructor at Trident Technical College in Charleston, South Carolina. Phil has contributed immeasurably to an enhanced awareness of the need for introductory engineering curricula through his professional presentations and publications.

I thank the following reviewers of this edition for their helpful comments: Charles Cofer, Chattanooga State University; Paul Dykshoorn, Stark State College of Technology; George Fredericks, Northeast State Technical Community College; Aron Goykadosh, New York City Technical College; and Doris Osburn.

The staff of Prentice Hall provided me with the professional support needed to make this text a quality product. Thanks to all of them.

I welcome your feedback and suggestions. You can contact me at 75234.1563 @compuserve.com.

Contents

1 **Engineering Technology as a Career**   **1**

2 **Career Choices in the Engineering Technologies**   **27**

1

Engineering Technology as a Career

All of us wish to understand the world we live in and to know our place in it. As children we depended on our parents to protect and feed us and establish an atmosphere we could depend on. In those early years, stability was necessary for normal development. As we became adolescents we experienced changes at school and with our friends. That unpredictable world was frightening and intimidating at first, but we learned to adapt. Most of us even learned to look forward to and expect change.

Change creates opportunities for those who prepare themselves with the skills, knowledge, and attitudes to solve problems. This book is about technology, and about the skills, knowledge, and attitudes possessed by the technologists who live and work in a world where "the only constant is change."

You may become a part of the exciting world of technology. To do so you will need to acquire a practical knowledge of mathematics and science. You must also learn how to communicate well with others. But, above all, you must be prepared to constantly adapt to the ever-changing world of technology.

Technologists are responsible for providing the material things necessary for human subsistence and comfort. Automobiles, transportation systems, buildings we work in, efficiently automated industrial and business processes, improved power systems, new materials, more powerful computers and highly integrated communication systems (Figure 1.1), are but a few of the commodities we expect from technologists.

Technology has improved our lives. At the turn of the century, half of the population of the United States lived on farms working from sunup to sundown merely to feed everyone. Now, with advanced farming techniques made possible by our modern technology, only 3 percent of the population feeds the entire United States plus a significant proportion of the rest of the world. And, while today's farmer may still work from sunrise to sunset, the work is much less labor-intensive.

Only since 1950 was the first commercially available computer born—HAL's great-great-grandfather. People first visited space. The first efforts at providing

Figure 1.1   Technicians installed the elaborate communication systems of the 1996 Atlanta Olympics. (Courtesy of International Business Machines Corporation. Unauthorized use not permitted.)

communications for the masses—e.g., direct long-distance dialing—was introduced. The first fax machine standards allowed different brands of machines to communicate.

People live much better than they did just twenty-five years ago. Consider the following: The number of autos per person has increased from 61 percent in 1970 to 73 percent in 1991. Gas mileage has more than doubled, while newly developed safety accessories like air bags and antilock brakes have been added. The average new home grew by 300 square feet between 1970 and 1992. An increasing number of the newer homes have central air and labor-saving devices like dishwashers, washing machines, and dryers. The price of a television set was $530 in 1970, compared with today's average price of less than $250. Home videocassette recorders didn't even exist in 1970, and now two-thirds of all households have a VCR. Other innovations in the last 25 years include compact disk players, microwave ovens, cellular phones, fax and answering machines, and the personal computer. The personal computer had 4000 bytes of memory in the 1970s, while today's memory usually exceeds 8,000,000 bytes. The internet (developed in the 1960s) and the World Wide Web (a new interface introduced in 1981) allow each of us to access information that only a privileged few had access to before their introduction.

Technology offers the world unparalleled opportunities. This textbook will help you better understand the world of technology while discovering how you might best step out into that world and enjoy an exciting and profitable career.

Figure 1.2 This Mayan pyramid (Yucatan Peninsula, Mexico) was precisely oriented to form the shadow of the seven triangles of the serpent's back only on the vernal and autumnal equinoxes, which occur in spring and fall respectively.

1.1 History of Engineering and Technology

Ancestral Engineering—Humankind's Search for Identity

To fully appreciate the world of technology we begin with some early history. The technologists and craftspeople of early civilizations built huge objects. The Great Wall of China was built by those who learned through trial and error. But its construction also required precise surveying and an amazing talent to use the lever and the inclined plane. Algebra and trigonometry were well understood and applied during those early years. Construction of the pyramids of Egypt and of Central and South America required experience (trial and error) and the labor of many people. In addition, however, many of the pyramids are oriented with great accuracy to the movement of the sun (Figure 1.2) or to the cardinal points of the compass. Such accurate positioning required the use of a well-developed system of mathematics and science. Sophisticated long-range planning was necessary in all of the great, early projects.

These huge constructions, so precisely located, helped humankind establish identity and satisfied the basic need to build and create. Engineering and technology activities satisfy this same basic need. The early builders were the forerunners of today's civil, mechanical, and mining engineers.

The Five Main Branches of Engineering (1700–Present)

Modern engineering and technology began in the 1700s and developed into five main branches: civil, mechanical, mining and metallurgical, chemical, and electrical.

Civil Engineering

The civil engineer, the earliest defined engineer, is the builder of our infrastructure—our foundation. You cannot have civilization without civil engineers, says the American Society of Civil Engineers. Some essential elements are public utility systems, buildings, roads, railways, airports, bridges, and waterways. Civil engineers must understand soil consistencies so they can design and build sound foundations. They must be familiar with the many types of construction materials and be able to determine the capabilities and limitations of each. Because they make structures that humans depend on, they must accept a high level of responsibility for their actions.

Future challenges for technologists in the field of civil engineering include modernizing our present infrastructure, building new systems to clean and maintain our environment, and developing more efficient power-delivery systems.

Mechanical Engineering

Steam power in the early 1800s brought the need for a new engineer, the mechanical engineer. Mechanical engineers and technologists made possible the development of machine tooling and manufacturing. Today's modern, automated industries are largely the result of the early efforts of the mechanical engineer.

Today's technologists in the mechanical area must understand energy-transfer and energy-conversion devices. Lasers (the acronym LASER stands for *l*ight *a*mplification by *s*timulated *e*mission of *r*adiation), gasoline engines, motors and generators, and fluid-power systems are but a few examples of such systems. Mechanical technology workers design, build, and test all types of machinery and work in most industries.

The aerospace industry is one of the challenges for mechanical engineering personnel. Improved materials must be used to build faster, more efficient, and safer aircraft. Space stations and space transportation vehicles are urgently needed for society to explore our solar system and to enjoy the rich resources of other worlds. Additionally, some manufacturing processes can be greatly improved in a gravity-free environment.

Mining and Metallurgical Engineering

The 1800s also saw the evolution of mining engineering. The need for coal to heat homes and fuel factories came first. Later, petroleum exploration and refining became necessary.

The mining engineer is concerned first with how to extract minerals safely and efficiently. Mining engineers and technologists must be familiar with civil engineering in order to construct safe, well-ventilated mine shafts. They must be aware also of soil conditions and related problems, such as drainage, making geology an important part of their preparation.

The metallurgical engineer is also concerned with extracting metal-yielding minerals from the ground. More often, however, this technical area is concerned with how to mold, cast, and shape metals and how to improve such metallic properties as strength, hardness, or stiffness.

Mining engineers and metallurgists face many challenges. Oil and mineral deposits below the surface of the ocean have scarcely been touched. Space vehicles require new materials to reduce weight, improve strength, and increase heat-dissipation properties. New ways must be found to protect the environment during and after mining.

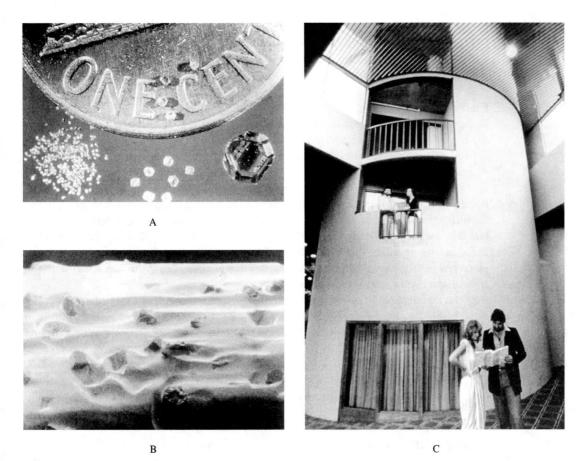

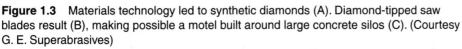

Figure 1.3 Materials technology led to synthetic diamonds (A). Diamond-tipped saw blades result (B), making possible a motel built around large concrete silos (C). (Courtesy G. E. Superabrasives)

Materials technology, closely related to the discipline of metallurgy, will bring important new materials to our homes, businesses, and industries (Figure 1.3). New uses for ceramics and polymers will improve our living standards.

Chemical Engineering

Chemical engineering evolved later in the 1800s, with society's increasing need for mass-produced chemicals. Like the metallurgist and materials technologist, the chemical engineer controls the chemical processes that convert raw materials into useful commodities.

Chemical engineers and technologists are specifically involved with the manufacture of chemicals. They are responsible for such varied chemically related industries as food processing, drug manufacturing, environmental control, and nuclear energy.

Future challenges in chemical technology will be in the continued search for more efficient production of pharmaceuticals, while maintaining the high quality already present in such processes, and improved testing of food products. The chemical engineer's greatest contribution, however, will be in the development of new energy systems. Fusion energy systems and improved solar energy systems will bring clean, safe, and cheap energy to homes and industries. Development of less expensive and nonpolluting energy systems is probably one of our society's highest priorities.

Electrical Engineering

Electrical engineering was not a distinct field until the 1900s. Young as it is, it is now the largest branch of engineering. Early in the century, electrical engineers were concerned primarily with the production and distribution of electrical energy—power plants and power lines.

Early electrical engineers worked in the mechanical world as much as the electrical. They were the first electromechanical engineering technologists. Charles Kettering, working with a group of highly creative technicians, invented the automobile self-starter in the early part of the century (see Chapter 3). Kettering and the "barn gang" are outstanding examples of the flexible, cross-disciplinary character of early electrical technologists. These versatile builders of yesterday contrast greatly with the specialized electronics engineers of today.

Electronics technology arrived with the development in 1904 of the vacuum tube, leading to the first amplifier in 1907. From the 1920s to the 1950s the vacuum tube led to the inventions of television and computers. In the 1950s the transistor and other solid-state devices replaced the energy-hungry vacuum tube.

The electronics industry is one of the largest employers of engineers today. Electronics engineers are found in most other industries as well. Challenges in the field of electronics will exist primarily in the areas of computers, machine control, and improved communication systems.

Today's Engineering Fields

Today, demand has created over thirty different engineering fields from the original five main branches. Some of these new degree areas are

Aerospace	Electromechanical	Materials
Agricultural	Engineering science	Nuclear
Architectural	Environmental	Petroleum
Biomedical	Industrial	Systems
Computer	Marine	Welding

The Emerging Need for Technicians and Technologists

With the launching of *Sputnik* by the former Soviets on October 4, 1957, the need for more specialized and more scientific engineers became apparent. By this time technology had become quite complicated, and new space systems offered almost overwhelming challenges. Four-year engineering schools were funded to upgrade curriculums and to produce engineering graduates with greater scientific skills.

Graduates of the engineering programs of the 1960s had more theoretical knowledge, but less practical knowledge and manufacturing experience. These new "engineering-scientists" achieved beyond expectations and allowed the United States to enter and win the space race of the 1960s and 1970s. New space systems planned for the twenty-first century promise to make space and other planets as much a home for humankind as earth is now. However, the practical engineer who could build and maintain traditional industrial systems became a rare commodity. Technical education was developed to bridge the gap. Technicians and technologists, graduates of technical programs, took responsibility for the more practical and less specialized scientific work.

1.2 The Industrial Team

Today's industries are divided into two fundamentally different types: manufacturing and service industries. Manufacturing industries make products. The need for the technician was first recognized in manufacturing.

Service industries provide services. A great number of technicians are needed in the high-growth service sector. Technologists in the service industries are needed to properly connect and repair the increasingly complex equipment used in our homes and offices. Most of this equipment utilizes computers, and computer-service technologists are in strong demand. Other technologists are needed to make service industries more efficient. Those engineering technologists will play an increasing role in the areas of quality control, supervision, and sales.

In both manufacturing and service sectors teamwork will be required if companies wish to survive. Teamwork in manufacturing means that the scientist, engineer, accountant, technician, technologist, and the skilled worker all cooperate in bringing improvements to manufacturing processes that produce goods. Teamwork in service means that the owner or manager of the organization trains and supports technologists to achieve customer satisfaction. The computer-service technologist sent to a customer's computer facility to solve a problem that has resulted in down-time will be under extreme pressure to fix the problem in a very short time. With the proper resources available from his or her company, the knowledgeable and well-trained technologist will not only satisfy but delight the customer. The result will be increased business for the service company.

The Role of the Technician and Technologist

The technician and technologist work in key positions on the industrial team. Persons holding two-year (technician-level) and four-year (technologist-level) degrees act as

1. communicators, providing clear communication between engineer and skilled worker;
2. implementors, interpreting the ideas of the engineer and implementing them;
3. calibrators and testers, performing complicated tests in engineering laboratories; and
4. manufacturing engineers, supervising skilled and semiskilled personnel and solving problems in manufacturing processes.

The technician acts as *communicator,* illustrating complex technical ideas so they may be understood by others in the workforce. This task is often accomplished by the preparation of engineering drawings or charts and graphs, and by direct communication

A   B

Figure 1.4   The design equations of the engineer and scientist (A) become reality with the work of the technician and technologist (B). (A, courtesy Hewlett-Packard; B, courtesy National Institute of Standards and Technology)

with the skilled employee. The technologist is responsible for planning and supervising. The supervisor requires excellent interpersonal communication skills, which include *listening* as well as speaking and writing skills.

For example, the civil engineering technician takes the plans of the civil engineer and prepares detailed drawings of a certain part of the project. The technologist then takes the drawings to the bridge or highway being constructed and directly supervises the construction personnel in the field.

Technicians and technologists must be able to speak clearly and accurately to enjoy credibility with coworkers and managers. They must not use confusing language when discussing detailed factual material.

Technicians and technologists take on active roles in *implementation*. This frees engineers to continue the flow of creative design ideas (Figure 1.4A) and to deal with broad concerns such as the personnel, managerial, and economic consequences of a project. Technicians implement the engineer's ideas, making the ideas reality (Figure 1.4B). They measure the quality of production, install new equipment, interpret the chart

Figure 1.5 Technicians and technologists repair, maintain, and calibrate sensitive measurement equipment. (Courtesy AT&T Network Systems)

recordings and gauges monitoring a manufacturing process, or supervise the construction of the superstructure (steel skeleton) of a large office building.

In research and development (R & D), technicians work with engineers to introduce new materials and processes and to test new materials for such qualities as strength and durability.

Calibration of test equipment is vital in today's industry. Heat sensors, flow meters, fluid-pressure sensors, and electrical measuring equipment such as oscilloscopes and voltmeters are a few of the many types of instruments used to measure industrial processes. The technician, trained to read schematic drawings of industrial measuring instruments and possessing the knowledge of how the instruments operate, must often repair, maintain, and calibrate them. The technician in Figure 1.5 repairs and calibrates communication equipment, ensuring reliable, accurate operation.

Technologists are today's *manufacturing engineers.* The typical four-year engineering technology curriculum provides a background equivalent to that of the baccalaureate engineer's curriculum of the 1950s. The technologist is often more willing and better suited to be involved with the day-to-day problems of manufacturing than is today's more scientifically educated engineer.

Figure 1.6 Engineers, technicians, and skilled workers team up to solve a manufacturing problem. (Courtesy AT&T Network Systems)

For example, a fiberglass insulation line slows because of the material sticking to the rollers. The supervisor of manufacturing engineering assigns the problem to a technologist. The technologist first analyzes the problem in light of the physical characteristics of the material, a task involving an applied knowledge of chemistry and physics. The technologist measures the temperature and humidity with a sling psychrometer and determines that humidity is higher than normal. Skilled workers are then assigned to inspect the duct system and find that a large ventilating fan is inoperative. The technologist then reports the problem and recommends the fan be replaced.

The preceding discussion illustrates the four essential services expected of technicians and technologists: communicator, implementor, calibrator and tester, and manufacturing engineer. By studying to become a technician in two years and perhaps continuing for a four-year bachelor of science in engineering technology (B.S.E.T.) degree, you will be well prepared to find a position in industry that will be both challenging and rewarding for years to come.

Teamwork in Manufacturing

Manufacturing industries require the teamwork of the scientist, engineer, technician and technologist, and skilled worker (Figure 1.6). The *scientist* is engaged principally in research and the development of new materials—in advancing the state of the art.

Table 1.1   The Industrial Team—Duties and Education

Duties	%Theory	% Applied	Education Required/Degree
Scientist—hypothesizes and verifies laws of nature	90	10	Five to seven years of college M.S. or Ph.D.
Engineer—designs and creates hardware and software from scientific ideas and laws of nature	70	30	Four or five years of college B.S. or M.S.
Technologist—makes design prototype, suggests redesign or modification, acts as manufacturing engineer	60	40	Four years of college B.S.E.T.
Technician—makes model of prototype, tests and troubleshoots prototypes and hardware/software in actual production use, acts as manufacturing supervisor	50	50	Two years of college A.S.E.T.
Skilled worker (craftsperson)—produces parts (e.g., holding fixtures) from completed designs, installs and runs hardware	20	80	Four years of on-the-job training (OJT) and/or vocational high school High school diploma and training/experience

M .S. = Master of Science, Ph.D. = Doctor of Philosophy, B.S.E.T. = Bachelor of Science in Engineering Technology, A.S.E.T. = Associate of Science in Engineering Technology. See Section 3.1 for descriptions of A.S.E.T. and B.S.E.T.

The *engineer* provides system design and technical management. The *technician* and *technologist* provide the practical, hands-on, manufacturing expertise. The *skilled worker* operates and repairs specialized machinery. Table 1.1 depicts the amount of theory and applied knowledge needed for these five industrial classifications.

The skilled worker performs tasks requiring some mathematics and other theoretical knowledge, but relies mostly on hands-on experience. Examples of skilled workers are machinists and electronic assembly workers. Indispensable to production and used often in maintenance, the skilled workers must be included in the industrial team.

In industry, the roles of the engineering technician or technologist and of the engineer may be clarified by comparing and contrasting their occupational tasks. To illustrate, the electronic engineer (EE, or "double E") is responsible for *designing* new computer systems, *directing fabrication* of mainframe computer systems, and *developing applications* for new technologies such as laser diodes. Occupational tasks of an electronic engineering technician (EET) are similar to those of the EEs in some areas, but differ markedly in others. Some occupational tasks the EET is responsible for are *designing the interfacing circuit* for a computer circuit board, *supervising the assembly* of specific electronic equipment, and *breadboarding circuits* designed by the engineer.

Figure 1.7 Technical students couple hands-on experience with theory in the process control laboratory. (Courtesy of Trident Technical College)

The italicized words in the preceding paragraph are key words that aid in comparing and contrasting the duties of the engineer and engineering technician. The engineer *designs systems,* but the technician *designs individual circuits;* the engineer *directs fabrication of systems,* but the technician *supervises the assembly of specific electronic subsystems;* the engineer *develops new systems,* but the technician *breadboards such systems, troubleshoots and improves circuitry,* and *brings the system into the real world.*

Another difference between engineers and technicians is how quickly they must learn their jobs. Because technicians are prepared in college for hands-on (practical) applications as well as theoretical applications (Figure 1.7), employers expect the technician to learn a new process or to perform a new laboratory test in a very short time. Employers have admitted that they expect a technician to be fully productive within 30 to 60 days after beginning employment. On the other hand, an engineer might take up to a year on the job to become familiar with the manufacturing process and become fully productive.

How may the members of the manufacturing team work together? Consider, for example, a hydraulic pump that continues to overheat because of inadequate system design or defective parts (Figure 1.8). An engineering technician, prepared to deal with the real-world problems of industry, is called in. An electromechanical, mechanical, or fluid-power technician is trained to test pumps and know what conditions must be present for pumps to operate effectively. The technician quickly isolates the

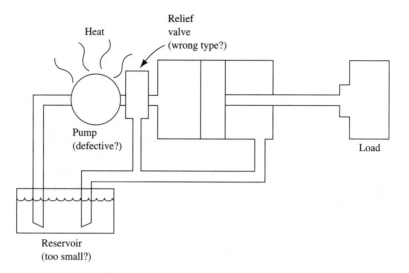

Figure 1.8   Training prepares a technician to troubleshoot a faulty hydraulic system.

problem by determining that (1) a specific type of relief valve incorporated into the inadequate system will solve the problem, (2) the oil reservoir is too small, or (3) the pump is defective.

   If new parts are required (e.g., oil reservoir, pump) the technician will order the substitutes by establishing the correct replacement specifications, researching suppliers' catalogs, and making telephone calls to determine availability. The skilled worker will complete the necessary maintenance by installing the new parts. Alternatively, repair of a part may be necessary. If the pump is determined to be defective, for instance, the skilled mechanic would be better qualified to tear down and rebuild the hydraulic pump than would the technician who has been trained only to test pumps, not to deal with the placement of seals and other internal components.

EXAMPLE 1.1

If the faulty hydraulic circuit in Figure 1.8 is diagnosed to require a different type of relief valve, which ports hydraulic fluid to the reservoir if system pressure becomes too great, who would specify the new relief valve? Who would determine the problem if complex piping or hose runs are at fault?

Solution   If the relief valve is at fault, the technician is quite capable of determining the specifications of the new valve and will specify a suitable replacement. If, on the other hand, the piping or hose (conductor) runs are long and complex, then the engineer is often needed to redesign the system and solve complicated fluid-mechanics problems. These problems may require sophisticated mathematics not possessed by the technician.

Communication and the Industrial Team

One of the critical roles technicians or technologists play is that of communicator. They "glue the industrial team together" and make it function as a whole (Figure 1.9). Technicians and technologists are best prepared to communicate with the skilled worker because of a technical education that includes a great deal of laboratory experience, including work with the actual tools and machinery used in industry.

During college, technicians and technologists are exposed to the language of the engineer, who uses information from mathematics, physics, chemistry, and other sciences to solve theoretical problems and to design new systems. The technician's college education requires a sound base of algebra and trigonometry (sometimes calculus) and at least two courses in physics or chemistry. The technologist's curriculum includes mathematics through calculus and advanced applied science courses.

Knowledge of the symbols and words used by both the skilled worker and the engineer enables the technician or technologist to become a critical link between them. Forging this link between the quite different worlds of the skilled worker and engineer is challenging. How do technologists build bridges (see Figure 1.9) and assure communication occurs between people who are expected to perform markedly different work? Technologists can accomplish these tasks only if they

1. have confidence in the real skills and knowledge they possess,
2. can logically and reasonably transmit their messages to others, and
3. know and can use the appropriate conventions of the language—for instance, good spelling, good grammar, and good sentence and paragraph structure.

Of course, *good human relations skills must be added to all of the above criteria.* One critical factor in human relations is listening to others. The technologist who can

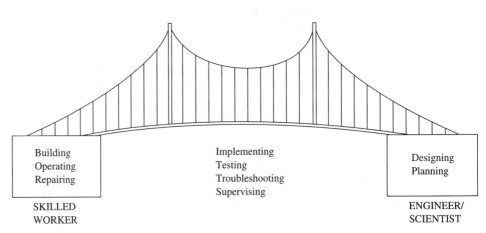

TECHNOLOGISTS AND TECHNICIANS

Building Operating Repairing	Implementing Testing Troubleshooting Supervising	Designing Planning
SKILLED WORKER		ENGINEER/ SCIENTIST

Figure 1.9 The technician acts as the bridge for the industrial team.

listen to the skilled worker will be much more effective and will learn a great deal more than those who can only direct others.

Organizational Structures and the Industrial Team

Medium to large companies must have an organizational plan for the industrial team to function correctly. The organizational plan fixes decision-making responsibilities, showing specifically how the members of the team interact and who makes decisions at a particular level. The plan may be shown as an *organizational diagram* or *organizational chart.*

Technicians and technologists should be aware of both formal and informal decision-making structures in their organization. Understanding the concepts of *line* and *staff functions* from an organizational chart is a good first step.

For instance, consider Figures 1.10A and 1.10B. In Figure 1.10A, the quality assurance manager is in a staff position, reporting to the plant manager and coordinating quality assurance activities between the group managers. The quality assurance manager acts only in the capacity of an advisor and does not enjoy line authority. In Figure 1.10B, the assistant plant manager is in a line position, no longer simply an advisor, but in a position that may involve directing the group managers.

Confusing line and staff functions may result in serious misunderstandings in an organization. It pays for the organization to be clear about who manages whom and for each employee to be clear about whom to report to. Figure 1.11 is the organizational structure for a large manufacturing company. Can you identify the line and staff positions in the diagram?

If you are a member of a staff organization (many industrial engineering functions are staff organizations), you must be prepared to work differently with others in the organization. A staff department must work with, not direct, other managers. Top management should assist staff organizations to do their jobs by breaking down any barriers between departments.

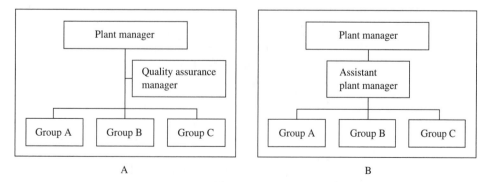

Figure 1.10   Staff (A) vs. line (B) relationships in an organization.

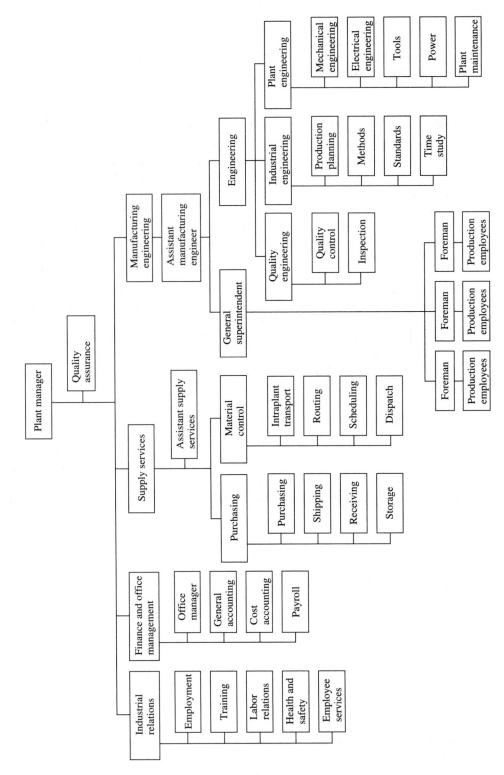

Figure 1.11 A typical organizational chart for a large manufacturer.

Competition vs. Teamwork

Competition in manufacturing or service industries should be directed toward competitors and not between members of the same organization. Top management has a central role in setting the climate for eliminating unhealthy internal competition. Companies where there is bickering and backbiting between employees and between departments will not survive in today's global economy. Such an environment is too inefficient.

Teamwork is the business strategy of the twenty-first century. Teamwork is enhanced by freeing teams to make more decisions on their own. Employees at all levels working in teams that consist of people from many disciplines make decisions on equipment needs, production levels, staff support, and even financial support for a particular project. In this environment, engineers and technologists interested in simply being left alone to work with a certain piece of equipment cannot survive long. They must learn to work with others.

Your courses in psychology and sociology will be of great benefit to you in learning how to work with others. In your technical laboratory courses you will be expected to work with at least one other person. Your technical instructors will assign real world projects that involve several students. Strive to understand others and your career will be greatly enhanced.

Service industries especially are looking for technologists who are flexible in dealing with customers. This means an ability to work with any client. Nontechnical clients are often put off by service technicians who can only use technical abbreviations and acronyms to describe what they do. Technologists must learn to explain what they do in popular terminology.

The greater the teamwork and the more personable the employees, the better will be the product or service. Technologists who can show they are willing to work with other people will be far more successful in obtaining a job. Companies used to hire on the basis of good grades. Now interviews are used to select graduates who can work with a team, have good communication skills, and have experience working on projects with others.

1.3 The Career Decision

To achieve the challenging and rewarding position of technician or technologist, you must first be convinced of the appropriateness of your goal. By knowing that becoming an engineering technician or technologist is the right career path, you will successfully complete the rigorous curriculum requirements of a college-level, paraprofessional program.

What are the important elements in career exploration? There are at least three: (1) occupational satisfaction, (2) availability of employment, and (3) salary potential.

Occupational satisfaction involves answering the question, "Will my day-to-day occupational tasks be enjoyable?" An older student, already experienced in industry, can satisfactorily answer this question. Recent high school graduates and others lacking industrial experience must base their decisions on personal experiences relating to the tasks an engineering technician must perform. The role of the technician in industry has already been examined. With that information you can address the following questions:

1. Do I enjoy working with equipment and machinery?
2. Do I enjoy working with numbers (data)?
3. Do I enjoy math and the sciences, especially when I can see how they apply directly to the real world?
4. Do I enjoy working with a group of people to achieve a common goal?
5. Do I enjoy communicating with others?

These important questions may be difficult to answer honestly at first. If more than one of the above questions receives a "no" answer, you should reconsider career goals.

One of the latest methods to determine your best career choices is to use a *career information delivery (CID)* system. A CID system is a computer-assisted career planning tool that enables the user to search through large amounts of occupational information at the click of a button. It works like a greatly expanded version of the questioning exercise above.

Discover is a popular CID system in use at many colleges (Figure 1.12). Students can explore what occupations offer the best employment opportunities, along with typical salaries, working conditions, and amount of education required. Or they may elect to interact with the computer and answer questions dealing with values, interests, skills, aptitudes, or experiences as they relate to work. *Discover* cannot give you magic answers or make decisions for you. If you still have questions about your career goals after using a CID system, consult an experienced guidance counselor or academic advisor.

It is easier to answer the second element in career exploration—*availability of employment.* Since the launching of *Sputnik* by the former Soviets in 1957 and the introduction of the computer, the need for technicians has exceeded the number of individuals capable of performing in this role. In short, there are employment opportunities available to engineering technicians and technologists.

The most recent *Occupational Outlook Handbook* may be found in your college library. The 1996–97 *Handbook* reports that well-qualified engineering technicians should find good employment opportunities through the year 2005. This forecast is due to the increasing need for technical products and services. The *Handbook* also identifies related occupations and projected openings for those classifications. Additionally, you may wish to consult the *Occupational Outlook Quarterly.* This periodical, issued every three months, can keep you even more up-to-date on expanding technologies and job prospects.

Automation is the key reason cited for optimistic employment projections in manufacturing industries. Each time the minimum wage set by Congress is increased, industry increases automation. Foreign competition has also increased the need for automation. Automation leads to better quality products by decreasing the variability in the parts of that product. Increased automation also adds to each employee's pro-

Figure 1.12 Students use the *Discover* program for career information (Courtesy Betty A. Proctor, Chattanooga Technical Community College)

ductivity, thereby decreasing manufacturing costs. The result is fewer employment opportunities for the semiskilled worker, but more employment opportunities for the technician and technologist (Figure 1.13A). The technologist is trained to consistently and reliably install and maintain automated manufacturing equipment (Figure 1.13B).

The recent advent of the microprocessor (Chapters 8 and 9) has enabled manufacturing industries to automate faster than before. This fact, coupled with more rapid machine processing and the use of improved materials in manufacturing (Chapter 10), has assured the increased need for technicians through the beginning of the twenty-first century.

The microprocessor has also invaded the burgeoning service sector. In the future, service industries will experience much greater growth than manufacturing industries, and salaries are improving. A recent Labor Department study shows that most of the 4 million jobs added from 1988 to 1993 were in relatively high-wage occupations paying $495 to $635 per week. Representative technical occupations

Skill rating

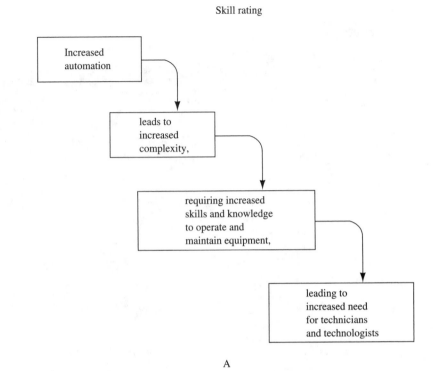

A

B

Figure 1.13 (A) Phases in the increasing need for technicians and technologists in industry. (B) Technicians and technologists maintain automated systems. (Courtesy of Trident Technical College)

Table 1.2 Salaries for Engineering Technicians

Experience Level	Per Year	Per Week
Less than one year*	$21,400	$410
With Experience**	$34,530	$660
Supervisory or Senior Level	$51,060	$990

* Abbot, Langer & Associates, Compensation and Benefits in Research and Development, 1995.
** From *Occupational Outlook Handbook,* 1996-1997 edition.

in the service sector include repairing computer, office machine, and electronic home entertainment systems; installing new software and networking office computers; and providing consulting services for the appropriate installation and use of new technical products.

National, state, and city governments hire *health and regulatory inspectors* to enforce a wide range of regulations that protect public health and safety. Employers prefer applicants with college training, but requirements include a combination of education, experience, and a passing grade on a written examination. Some examples of specific jobs include consumer safety inspectors, occupational safety and health inspectors, sanitarians, and pollution control engineers.

Your technical college will have information regarding local employment projections for the particular technologies offered on your campus. Responsible colleges offer only those technical programs that prepare graduates for positions needed now and in the future.

What about *salary potential?* Technical students should know the salary to expect after completion of an engineering technology program. Typical salaries for engineering technicians are presented in Table 1.2.

Table 1.2 shows that *starting salaries are not as important as the long-range earnings for the technical employee.* For instance, the table gives the annual earnings of engineering technicians having less than one year of experience as $21,400 in 1994. But, those with experience in 1993 earned $34,530. The most successful technicians, Supervisory or Senior Level, earned still more.

Technologists are usually hired as applied engineers and command even better salaries. According to the *Occupational Outlook Handbook,* 1996–1997 edition, beginning engineers had median annual earnings of about $34,000. Experienced mid-level engineers with no supervisory responsibilities had median annual earnings of $54,400. Senior-level engineers made $90,000 per year!

Check with your local college placement office concerning earnings of graduates. Local information is more useful than national averages.

EXAMPLE 1.2

You have completed your associate degree and are planning to continue in college for two more years and graduate as an engineering technologist. Your instructor gives you a rule of thumb of 50 percent as the increase in starting salary over the technician starting salary in Table 1.2. Calculate your anticipated annual starting salary as a technologist.

Solution From Table 1.2 the starting salary for a Junior-Level Technician is $21,400 per year. The annual starting salary is multiplied by 1.50, automatically adding the 50 percent figure to the base salary.

$$\$21,400 \times 1.50 = \$32,102 \text{ (rounded)}$$

The aspiring technologist can expect a starting salary above $30,000 per year. Technologists require two additional years of formal education, consisting of more mathematics and science, greater depth in their chosen technology, and management training. They can usually expect a starting salary of 50 percent or more than that of the technician.

A national report, *What's it Worth? Educational Background and Economic Status: Spring 1984,* found a clear relationship between salary and the technician with or without an associate degree. *The technician with the associate degree averaged 15 percent higher salary than one who had attended college, but left without earning a degree.*

The 1984 report has been continued into the 1990s and the findings are that "Most degrees beyond high school have significantly higher earnings associated with them than the next lower degree." People in the workforce in 1992 had the following projected lifetime earnings:

- $821,000 with a high school diploma
- $1,062,000 with an associate degree
- $1,421,000 with a bachelor's degree

The estimates are for all degree fields. Engineering and technical workers may enjoy larger earnings. The forgone conclusion is, **stay in school and earn more money!** (See Figure 1.14)

One way for industry to assure an adequate number of technicians and technologists in the future is to recruit more women. Women have become a necessary part of the entire workforce; after all, they represent approximately 50 percent of the population. In the technologies especially, there are too few women candidates. Colleges are attempting to alleviate the problem by attracting more women to their programs. Industry is responding, with encouragement through equal employment opportunity (EEO) legislation, by giving preference to females adequately trained to fill manufacturing positions. Female technologists also earn more than other females in the workforce. Female technologists earn 84 percent of what male technologists earn. Other females, in nontechnical jobs, earn only 69 percent of male earnings. A female considering a career in the engineering technologies is definitely on the right track.

Other minorities are also being sought as engineering technicians and technologists. All of the reasons cited—need for more technicians and technologists, lack of minorities in technology positions, and EEO legislation—may be used to show why all minorities (regardless of race, color, sex, disability, and age) should consider the engineering technologies. Well-qualified minority engineering technicians have had excellent employment opportunities in the recent past. For instance, in a 10-year period (1984–94), the number of electronic technicians of black or Hispanic origin increased

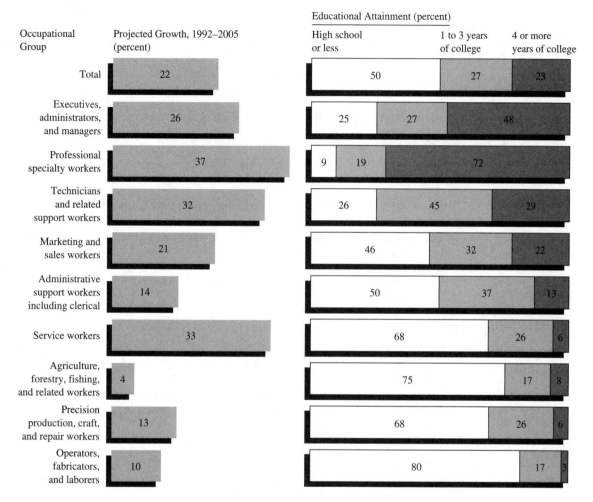

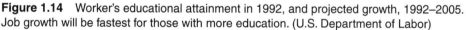

Figure 1.14 Worker's educational attainment in 1992, and projected growth, 1992–2005. Job growth will be fastest for those with more education. (U.S. Department of Labor)

by 80 percent. In the service sector, the number of black and Hispanic electronic repairers almost quadrupled in the same 10-year period.

People with physical disabilities will also find an increasing number of positions in industry. Automation makes their skills as valuable as the skills of those without physical disabilities.

1.4 Engineering Technology—The Right Career for You

This chapter has provided you with career information that will help you decide whether to pursue a technical degree. As previously discussed, engineers are charged with the initial design of systems; they must make managerial and economic decisions. But the technician and technologist must make the plans work.

With the proper background and training, you can become this implementor. You will draft or build a model of the new product, set up the new assembly line, install new production equipment, or supervise the construction of bridges or office buildings. You must be able to answer questions such as the following:

1. What product materials can be used and what is available?
2. What drawings must be completed and what conventions or specifications must be followed?
3. What machinery must be used to complete the task?
4. How can the work be supervised most efficiently?

If you enjoy being challenged and can work with a team to accomplish a goal, you will enjoy your career as an engineering technician. Rewards such as good salary potential and excellent employment opportunities can make engineering technology the right career for you.

Problems—Chapter 1

Section 1.1

1. Research and select one technological development that has occurred during your lifetime and list at least two ways it has improved your life. Also, list at least one potentially harmful impact it may have on your life and the lives of others.

2. Pick one of the original five main branches of engineering. Through personal research discover one early contributor to the selected branch and describe his or her contribution in two paragraphs or less.

3. List four technical devices or appliances, *in common use* today, that were unavailable in 1950.

4. Interview a person aged 60 or older. List what he or she feels has been the most significant technological developments in his or her lifetime.

Section 1.2

5. In one paragraph, discuss whether you would like to be employed in a service or manufacturing industry. Support your answer in a second paragraph.

6. In four or five sentences, describe personal qualities you possess that would lend themselves to effective teamwork on a job.

7. List four services the technician or technologist is expected to perform in industry.

8. Describe in three sentences or more how the role of the technician/technologist compares with the role of the engineer in manufacturing.

9. Through personal research with friends, neighbors, and other acquaintances, surmise and diagram the organizational chart for a large company that exists in your area or a

local company selected by your instructor. Compare your hypothetical organizational chart with other students' charts.

10. For the following listed job functions, classify each as most appropriate for the technician or technologist (1), skilled worker (2), or engineer (3). See Example 1.1 and the first job function with a correct response.

Job Function	Classification (1, 2, or 3)
a. repairs a hydraulic control valve	2
b. designs a new computer	
c. supervises an assembly operation	
d. troubleshoots a faulty hydraulic circuit	
e. designs a bridge	
f. repairs a T.V. set	
g. breadboards (constructs) a newly designed circuit	
h. designs private dwelling and submits plans to a Registered Architect for approval	
i. consults on materials for a skyscraper	
j. designs an elevator	
k. supervises the installation of elevators	

Section 1.3

11. List three crucial elements you should consider when deciding on a career.

12. Use the career information delivery (CID) system in your library or career services center. Submit your findings as to whether you feel a career in engineering technology is appropriate for you.

13. What does an increase in automation (e.g., robotics and computer-aided manufacturing) have to do with the need for technicians in industry?

14. Locate the *Occupational Outlook Handbook* in the college library's reference section. Research the projected number of job openings in three job classifications of interest to you. Organize and submit your findings to the instructor.

15. Locate one issue of the *Occupational Outlook Quarterly* containing information on openings for technicians in the college library's periodical section. Research the projected number of job openings in a job classification of interest to you. Organize and submit your findings to the instructor.

16. You have just graduated from a two-year program and are considering a four-year degree program in the engineering technologies. Assume a 60 percent increase in starting salary as a technologist over that of a technician. Calculate your estimated starting salary with the four-year degree using the method shown in Example 1.2.

17. Prepare a two- to three-paragraph report on the content of Figure 1.14. In the report describe how *you feel* concerning job availability and earnings as related to level of education.

Section 1.4

18. Design a job questionnaire. It should include such questions as What do you enjoy most about your work? What do you enjoy least about your work? How do you interact with engineers and skilled workers on the job? What technical skills and knowledge (e.g., math skills, drafting skills) are vital to your job? Which technical skills and knowledge are seldom used, if at all? What promotions have you already received and when? Add at least three other questions. Then interview a working engineering technician/technologist (preferably a degree holder). Submit the completed questionnaire to your instructor.

19. Access the internet from your school's computer system (see Section 8.4). Visit the following career information websites:

    HTTP://WWW.AJB.DMI.US

    HTTP://WWW.ESPAN.COM

    HTTP://WWW.JOBTRACK.COM

    HTTP://WWW.ASPENSIS.COM

    HTTP://WWW.QUESTMATCH.COM

    HTTP://JOBQUEST.COM/OOH1996/CONTENT1.HTM

    Download parts of a webpage to attach to a one-page report on your findings.

20. With a team of students organized from your class, find a new technology used in your community. Prepare a five-minute oral report for the class.

Selected Readings

Occupational Outlook Handbook, 1996–97 ed., Bulletin 2450. Washington, DC: U.S. Department of Labor and Bureau of Labor Statistics.

Occupational Outlook Quarterly. Washington, DC: U.S. Department of Labor and Bureau of Labor Statistics.

2

Career Choices in the Engineering Technologies

This chapter includes the major technical areas used to classify technicians and the responsibilities of those positions. If you do not find your chosen technology listed, it is probably included as a part of one of the following six general areas. Your instructor may modify this list according to the particular classifications used in your college's service area. The major areas are

1. chemical engineering technician (ChET),
2. civil and architectural engineering technician (CET),
3. electrical/electronic engineering technician (EET),
4. computer engineering technician (CpET),
5. industrial engineering technician (IET), and
6. mechanical engineering technician (MET).

2.1 Chemical Engineering Technician

Chemical engineering technicians (ChETs) generally work in three major areas of industry: (1) research and development (R&D), (2) production, and (3) technical sales. Chemical engineering technicians and technologists work with little or no supervision in a team composed of research scientists and chemical engineers.

In R & D, the ChET sets up and operates laboratory apparatus to test products for such characteristics as clarity, content of specific chemicals in the material, temperature sensitivity, and strength. The ChET uses complex instrumentation to measure the effects of temperature change on materials, and to collect strength and hardness data. Often, he or she uses a computer to analyze the data. The R & D area also includes small production lines (pilot lines) that allow technicians to prove the feasibility of a manufacturing plan or process.

EXAMPLE
2.1

Research a specific technology within the broader occupational area of Chemical Engineering Technology (ChET) that is expected to expand within the next decade.

Solution   The fastest growing industry in the whole economy in terms of output will be drug and pharmaceutical products, according to *Projections 2000,* a government source (Figure 2.1). According to the ***American Technical Education Association Journal*** (October–November 1995, page 21), several hundred experts agreed that biotechnology will grow faster than all other emerging technologies. Other emerging technologies making the top five in the study were materials (discussed in this section), electronics (Section 2.3), and computers and telecommunications (Section 2.4).

Environmental concerns about manufacturing have increased. The *environmental engineering technician* helps ensure that our air and water are protected. The Environmental Protection Agency (EPA) and other legislative bodies have recently enacted laws in defined areas such as air and water pollution, soil and groundwater pollution, toxic materials used in products (e.g., asbestos), and the use of pesticides. The environmental engineering technician's work is performed outside, collecting samples, and in the laboratory, analyzing the samples (Figure 2.1).

Figure 2.1   A biotechnologist collects data on biomaterials in the laboratory.

The *production technician* supervises or operates the manufacturing process in the plant and inspects the quality of the product or the amount of product produced per day. Often, the chemical engineering technologist directs skilled workers to adjust valves that regulate equipment, start pumps or compressors, or shut down a system if safety is questionable. (The 1979 incident at the Three Mile Island nuclear energy plant would not have occurred if operating personnel had been properly trained. At Chernobyl, operators were instructed to *disobey* standard operating procedures.) To verify process conditions, the technician bases most decisions on observing meters, gauges, or chart recorders.

Technical sales and development (TS&D) technicians are concerned with identifying and meeting the needs of the customer. They test and determine characteristics of the product (e.g., color, taste, or durability) that will have an effect on sales, then sell the product to distributors and train them on the benefits and limitations of the product.

You may find the following list helpful in understanding the broad range of specialization in chemical engineering technology. Each specialty is usually needed in a particular geographical area. In these areas technicians may be trained for the specialty field. For example, a nuclear engineering technician program may be offered at colleges located near nuclear power plants.

Materials engineering technology	Biomedical/biological engineering technology
Ceramic engineering (i.e., glass) technology	Nuclear engineering technology
Plastics engineering technology	Petroleum engineering technology
Metallurgical engineering technology	Environmental engineering technology

Nuclear engineering technology is usually considered a separate technology—not a part of the chemical engineering technology area. All of the above technologies may be offered as specific two-year or four-year technology programs at your college.

Many chemical engineering technologists will be involved with composite materials (Figure 2.2). Formed of a combination of glass and plastic, composite materials are unbreakable, lightweight, corrosion-free, and easily repaired. Composites can be manufactured with attractive surface finish and high strength and flexibility. Some desirable properties of composites that can be controlled are electrical conductivity, vibration, damping, spring rate, and tribology (a quality characteristic used in bearings, gears, and computer disk drives). Within the next decade, composite materials will be manufactured as cheaply and quickly as plastics. Composite usage will then increase dramatically. Composite material advances in the near future will make chemical engineering technology a challenging and rewarding career field.

A 1994 report shows chemical engineering technologists in R & D were paid more than any other technical discipline. According to the study, ChETs had a median income of $45,000. Projected employment growth through 2005 is moderate at 25 percent.

Personal qualities needed in order to achieve as a ChET include

- an ability to work well with others as part of a team
- an aptitude for science and mathematics
- accuracy and patience when conducting laboratory tests
- an ability to exercise care when working with toxic chemicals or disease-causing organisms

Figure 2.2 A technical associate in chemistry holds a photo produced by a computerized color imaging technique he invented. The result displays the elements of a composite sample. (Courtesy Lawrence Livermore National Laboratory)

2.2 Civil and Architectural Engineering Technician

Civil and architectural engineering technology is offered in many technical colleges as *construction engineering technology*. The *Occupational Outlook Quarterly*, Fall 1993, reports that "Construction will add almost 1.2 million jobs between 1992 and 2005, an increase of 26 percent, as it recovers from the residential and commercial building slump of 1991–92."

Civil engineering technicians and technologists (CETs) are perhaps the most versatile of all technical workers. They are responsible for such varied occupations as surveying roads (Figure 2.3A & B), implementing the plans for large structures, supervising the building of highways, constructing bridges (Figure 2.4), environmental engineering, and inspecting all parts of the infrastructure. *Property surveying is often*

A   B

Figure 2.3 (A) Civil engineering technicians survey a new highway. (Courtesy Hewlett-Packard) (B) A technical student trains with the latest surveying equipment known as the Total Station. (Courtesy Florence-Darlington Technical College)

Figure 2.4 Technologists produce and maintain commodities necessary for our existence. This bridge is being repaired by civil technologists. (Courtesy G. E. Superabrasives)

offered as a separate technical degree program. Property surveying technicians trace deeds, find and use old survey markers, and utilize special computational processes.

Increasing numbers of CETs should be required in the near future for two major reasons:

1. A large backlog of public-works rehabilitation and renovation projects exists in the United States. Such infrastructure repair projects include highways and bridges, sanitary systems, mass transit systems, waterways, and office buildings.
2. Internationally, the superior technology of the U.S. construction industry is being recognized. Increasing overseas employment could result (e.g., in *Stroyindustriya '87,* the director of the largest Soviet trade organization was quoted as saying, "Visitors to the U.S. exhibit at Stroyindustria's trade show have numbered around 150,000, with constant attendance by high-level delegations representing every ministry having anything to do with construction in the U.S.S.R.").

Architectural design technicians select appropriate building materials and build structures that are safe, attractive, and efficient. They consult on repairs, prepare final drawings for private dwellings, and confirm compliance with building codes.

They are often design originators as well as acting always as the design producers—private dwellings may be designed by those who are not Registered Architects (RAs). Architectural design technicians are also energy technicians. They are responsible for specifying heating and cooling systems that conserve rather than waste energy.

Many civil and architectural design projects involve the safety of large groups of people. Bridges and large buildings must be designed by the civil engineer approved for professional engineering (PE) or by an RA. Both civil and architectural technologists are the design implementors. They carry out the extensively researched and considered design of the professionally registered engineers. Concern for public safety and the need for improvements in construction quality will lead to faster than average employment growth for construction and building inspectors. Job prospects will be best for technologists with prior experience in construction industries.

The *Economic Research Institute* reported that the average starting salary was $23,000 in 1995 for a CET. The average salary was $27,500 for all working CETs, including new workers. Experienced workers earned $31,500.

Personal qualities needed in order to achieve as a CET include

- a willingness to work with others as part of a team and to direct others
- an aptitude for science and mathematics
- a need to exercise creativity and an ability to make decisions
- an ability to think and plan ahead
- a willingness to travel and work outdoors

Drafting and Design Technician

Drafting and design is central to many industries, as well as civil and architectural engineering technologies. Drafters produce the design drawings used to guide others who build structures or equipment. It is important to realize that their work is not simply to draw up the plans, but also to specify the right materials and establish suitable dimensions and tolerances. Drafters use handbooks, scientific calculators, and computers in their design work.

Manufacturing industries employ both electronic and mechanical drafting and design technicians. Electronics drafters must know the symbols and electrical concepts used to build suitable circuitry. Experienced electronic drafters are in demand and often command large salaries. Mechanical design technicians are discussed further in Section 2.6.

Drafting and design technicians using computer-aided design (CAD) systems are known as CAD technicians. The CAD system is a computer that allows designs drawn on a cathode ray tube (CRT) screen to be stored and easily modified later (Figure 2.5). For example, when designing a construction project, electrical and plumbing schematics may be layered on the system. Layering shows how the various elements will fit together and eliminates the necessity for many laborious drawings of the same building.

CAD must become more prevalent, and those who know how to effectively use such systems will remain competitive in the industry. It is especially important for

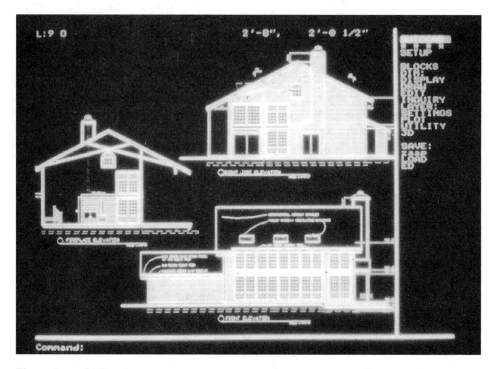

Figure 2.5 CAD systems are rapidly changing the construction industry. (Courtesy Autodesk)

students who aspire to owning their own contracting business to learn and integrate CAD systems in their operations.

Employment for both Drafting and Design and CAD Technicians is mixed. Because of the increased use of CAD systems, productivity increases may mean fewer jobs. However, many job openings are expected to arise as drafters are promoted, move on to other occupations, or retire. According to the *Occupational Outlook Handbook,* "Individuals who have at least 2 years of training in a technically strong drafting program and who have experience with CAD systems will have the best opportunities."

2.3 Electrical/Electronic Engineering Technician

Electrical engineering technicians install, control, and troubleshoot electric power distribution systems. They are almost always involved with high voltages and currents. This means that they must be able to interpret the National Electrical Code (NEC) and standard procedures established by the National Electrical Manufacturers Association (NEMA). Often, the electrical technologist will supervise skilled workers—electricians.

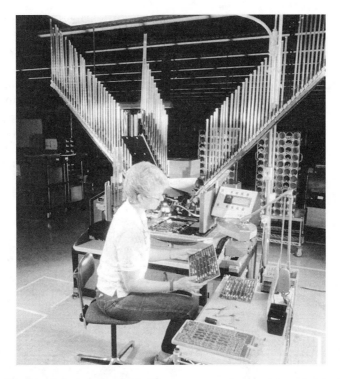

Figure 2.6 Electrical engineering technicians may require electromechanical skills to install and maintain machines such as this one, which performs programmed insertion of microchips and other components. (Courtesy AT&T)

Sound safety practices must be communicated clearly to all electrical workers, so they may protect themselves as well as others in contact with equipment they have installed. Even though electrical currents may be dangerous, the safety record for electrical workers is good.

Many electrical engineering technicians benefit from an electromechanical program. Industrial systems include motors, generators, switchgear, and other electromechanical devices (Figure 2.6). Electromechanical personnel are the maintenance personnel of industry. They are responsible for keeping complex automated systems operating properly. They also install and test automated systems.

The abbreviation EET is most often used for the electronic engineering technician. EETs help develop, manufacture, and service equipment such as audio and video systems, radar and sonar systems, industrial instrumentation systems, and medical equipment. The EET is usually involved with computers and may be appropriately known as a computer engineering technician (CpET). The rapidly changing computer area is covered in the next section.

The best employment growth for EETs will be for *service technicians* and *field service representatives.* They install, maintain, and repair electronic equipment used in homes, offices, factories, and hospitals. The best growth is forecast for *office machine repairers.* This occupation will grow faster than the average through the year 2005 as the number of computer-based machines increases (also refer to Section 2.4, Computer Service Technologists).

Perhaps one of the most rewarding long-range technical careers is instrumentation, which involves the control of manufacturing processes (Figure 2.7). *Instrumentation technicians* must understand such devices as electromechanical transducers that measure process characteristics such as heat and pressure, and the computer controller that keeps process characteristics under control. Instrumentation technicians work in sales, manufacturing, installation, and maintenance of all types of transducers and controllers.

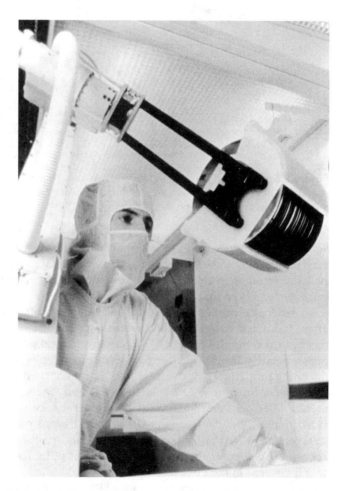

Figure 2.7 Instrumentation technicians control industrial processes. (Courtesy of International Business Machines Corporation. Unauthorized use not permitted.)

Service and instrumentation technicians also benefit from an electromechanical program. *Electromechanical technicians* have an understanding of hydraulic, pneumatic, and mechanical transfer equipment as well as electronic circuitry. In short, they understand electronics and mechanics. Automated machinery has electrical brains and mechanical muscle. A computer printer has both electrical and mechanical parts. The electromechanical technician is competitive in both manufacturing and service industries.

EXAMPLE 2.2

List two other occupations, other than those already listed, that are electromechanical occupations. What are employment growth prospects and salary ranges?

Solution   Two other occupations requiring electromechanical skills and knowledge are

- laser and fiber optics technicians
- biomedical equipment technicians

The employment outlook for electromechanical technicians is expected to be good through the year 2005. The number of electromechanical and biomedical equipment repairers is projected to grow 37 percent between 1992 and 2005. The average salary range is $28,000 to $33,000.

Other electronic engineering technicians work in one of two major areas: (1) communication electronics or (2) computer electronics. This section will consider the communication electronics technician; another section will be devoted to the computer electronics technician.

More than 24 percent of the 2.5 million workers in the U.S. electronics industry are employed in producing communications equipment. Many *communication electronics technicians* are employed in radio and television broadcasting industries. Recently, cable TV systems have enhanced employment opportunities in this area. In a small station the technician is classified as a *chief technician,* is diversified, and performs tasks such as specifying (deciding what types of equipment to buy), connecting, and maintaining broadcasting or cable equipment. This equipment includes microphones, recording equipment, transmitters, receivers, sound and lighting control systems, television cameras, antenna towers, satellite receiving stations (dishes), and signal-processing equipment. In a large broadcasting station the technician may be specialized and concentrate in one or a few of these areas. The chief technician ensures the system's compliance with all federal, state, and local safety regulations and usually is required to be certified by the Federal Communications Commission (FCC). Students interested in entering the broadcasting industry should check with their instructor to determine whether their college prepares them for FCC examinations.

The U.S. Bureau of Labor Statistics reports that employment of broadcast technicians is expected to grow only 4 percent through the year 2005. This low employment growth projection is due to labor-saving technical advances such as computer-controlled programming and remote control transmitters. Strong competition for jobs will continue in major metropolitan areas, and prospects for entry level jobs will be best in small cities.

A few ambitious communications technicians become entrepreneurs, building their own stations and managing them. The technician choosing this career should take courses in small business administration.

Closely related to communications electronics are radar and microwave. Both areas involve sophisticated electronic circuitry, and the technician must service these systems on a regular basis (perform preventive maintenance).

The *Economic Research Institute* reported that the average starting salary was $20,500 in 1995 for an electrical/electronics technician. The average salary was $34,000 for all workers, including new workers in this field. Experienced workers earned $63,500.

Personal qualities needed in order to achieve as an EET include

- a willingness to work with others as part of a team and to direct others
- above-average mathematical ability
- analytical ability
- mechanical and science aptitude

2.4 Computer Engineering Technician

The *computer engineering technician (CpET)* services, connects, and programs microcomputers or maintains the larger mainframe computer systems (Figure 2.8) used in both business and industry. In manufacturing, the microprocessor is the "brain" of the machine. In fact, the ability to automate in industry depends on the technician's ability to successfully interface (connect) the microcomputer to a specific machine or process.

Everyone is familiar with the rapid growth of the personal computer industry. Personal computers are now used extensively in business, industry, and homes. During the 1980s the proportion of industrial workers using computers increased from 18 to 28 percent. These computers must be serviced periodically by computer service technicians (often called field engineers or customer service engineers). *Computer service technologists* are usually assigned to several customers. This means they are often on the road. Also, they need strong public relations skills to deal with sometimes irate business customers who depend on their computers and suddenly find that the computers no longer work.

In all areas of computer troubleshooting and repair, the technician must be able to isolate the particular circuit board or, less likely, the malfunctioning component. Many times the diagnostic work can be done by the computer, but the technician must be able to understand the various computer operating systems and to use the system's software. Electromechanical skills and knowledge are important to the CpET in industry. Without some understanding of industrial processes,

Figure 2.8 Data communication technicians often act as troubleshooting analysts.

largely mechanical, the technician cannot appropriately program the computer to maintain the correct sequencing of operations, and avoid "crashing" the tool into the work (Figure 2.9).

An interesting technology, linking the skills and knowledge of the communications electronics technician with those of the CpET, is evolving. The *data communication technician* is responsible for building and maintaining specialized telecommunications equipment that transfers the enormous load of data that business and industry generate.

There is a great need for *network specialists.* It is one of the fastest growing occupations in the United States, second only to health care. As more companies depend on computer systems that are interconnected, trained network technologists will continue to be in great demand.

There is strong demand for computer technicians of all types. Companies report they are unable to fill all their openings in many cases. The job market for computer technicians will grow by 70 percent between 1994 and 2005. The U.S. Department of Labor lists computer and data processing services as the third fastest growing industry during this period.

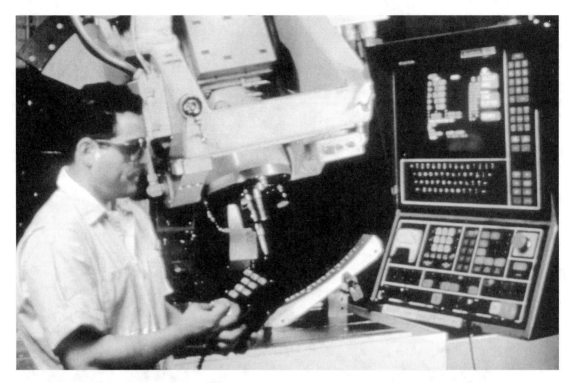

Figure 2.9 The CpET programs a machine. (Courtesy of International Business Machines Corporation. Unauthorized use not permitted.)

Personal qualities needed in order to achieve as a CpET include

- a willingness to work with others as part of a team
- problem-solving ability
- a willingness to work long hours to meet deadlines
- a constant drive to be retrained on new hardware and software
- a willingness to travel

2.5 Industrial Engineering Technician

Industrial engineering technicians (IETs) are involved in methods and time study, production control, quality control, and industrial supervision. This means they are generalists, working across many technologies, and are responsible for the integration of personnel, materials, machinery, methods, and plant layout.

The IET performs *methods and time study,* ensuring the efficient use of personnel and machines in a manufacturing operation. The tasks involved are studying and recording the time to do a particular job (recent changes in management philosophy—

Figure 2.10 Methods studies by industrial engineers and technicians result in a new automated materials distribution center for Scotland. (Courtesy International Business Machines Corporation. Unauthorized use not permitted.)

e.g., just in time (JIT) manufacturing—have decreased the need for time study), preparing charts and graphs illustrating work flow and efficiency, preparing layouts of machinery (Figure 2.10), specifying the number of machines necessary for an assembly line, and performing statistical analysis of work flow with computers.

It should be noted that methods studies are necessary in areas other than manufacturing. Service industries such as grocery stores, fast-food restaurants, and amusement parks employ industrial technicians to improve methods (Figure 2.11)—to show people how to "work smarter, not harder."

Production controllers use data from customer receipts and post-production records to set the production schedule for the plant. They may also measure the production of a particular manufacturing process. An important contribution is coordinating production between sections or divisions. Controllers move the product between sections and aid in decisions concerning the return of improperly manufactured items to the section responsible. They also inform the sales staff of the inventories ready for shipment to customers.

Figure 2.11 Service industries such as this retail store benefit from the knowledge of industrial engineering technicians. (Courtesy International Business Machines Corporation. Unauthorized use not permitted.)

One of the most important members of the manufacturing team is the *quality controller.* If the product of the plant is not a quality item, the plant cannot compete with other manufacturers and will be shut down. Quality controllers are involved in checking such product quality characteristics as

- dining room tables for a beautiful finish,
- automobiles for number of defects on leaving the assembly line (e.g., squeaks, rattles, scratches, poor fits),
- gears for runout (a measure of out-of-roundness),
- paint for its viscosity (its ability to spread easily), and
- restaurants for average time of service.

Checking the product's quality characteristics is called *inspection.* Many quality control personnel are responsible for measuring quality and for implementing the overall quality system in a plant—not simply for performing inspection. These technologists usually carry the title of *quality assurance (QA) specialists.*

In the 1980s an inspection technique called *statistical process control* (SPC) was introduced to many American industries for the first time. This recent adoption of SPC is due mainly to increasing foreign competition. SPC is a technology that moves an in-

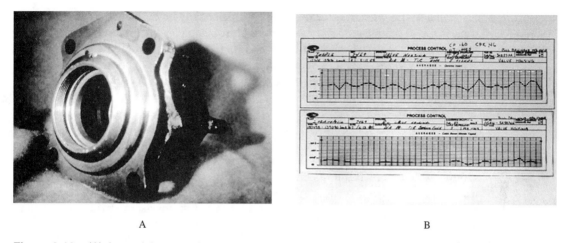

A   B

Figure 2.12 (A) A special composite material machine tool is used to machine this cast iron power steering valve housing. (B) A control (run) chart statistically controls the process. (Courtesy G. E. Superabrasives)

dustry from an inspection-after-the-fact quality control process to one of incorporating quality decisions in all phases of the product life cycle. SPC prevents quality problems from occurring. SPC also involves all of the workforce. Employees are no longer told to "just run the machine and don't think about what you are doing." Instead, they are trained to statistically understand the natural variation in all processes, to monitor these natural variations, and to identify variations that are not natural (that is, not due to their particular process).

By minimizing unnatural variations, a process may be brought into "statistical control." It is then predictable and repeatable and is capable of turning out one conforming part after another. Output from a process "in control" requires less inspection. In addition, abnormal variation can be spotted and corrected before the process drifts too far from a target value.

Many companies, especially in the auto industry, mandate that their suppliers adopt SPC methods. For instance, the manufacturer of the power steering component in Figure 2.12 could not supply major automobile assemblers unless SPC methods were in use. The major automobile companies also use statistical methods to check incoming materials. These "incoming inspection or acceptance checks" identify supplier problems before they affect the manufacturer's product. See Section 3.3, SPC Problem Solving.

Many business and office operations will also find that SPC can make them more competitive in today's world. Whatever the business or industry, to survive, it must evaluate and control product quality. Careers in quality control offer experiences in such diverse areas as developing quality control systems (planning), testing and inspecting products (implementation), encouraging and training others to improve and maintain quality (human resource development), and calculating and controlling costs (quality cost or savings studies).

Finally, the industrial engineering technician may be employed as an *industrial supervisor*. In manufacturing, supervisors must be able to understand the equipment and materials they are dealing with. In addition, they must communicate the ideas of management to the skilled or semiskilled worker. This makes labor-management relations an important course for the IET planning to enter supervision.

The *Handbook* states that blue-collar worker supervisors, "ensure that workers, equipment, and materials are used properly and efficiently to maximize productivity." Supervisors coordinate with maintenance to make sure that equipment is set up properly and is well maintained. Employee records and work schedules are maintained by the supervisor, who must be a caring and articulate person with sound management training. Employers are increasingly hiring college graduates with technical degrees to work as supervisors. Median weekly earnings for supervisors were about $610 in 1994. The highest-paid supervisors earned more than $1080 per week.

EXAMPLE 2.3

The service sector (nonmanufacturing) will grow by 30 percent by 2005. Explain how the IET can work in this burgeoning career area.

Solution   Industrial engineering technicians and technologists will be needed to work in occupations that distribute and keep track of the products delivered to consumers. They will be needed to maintain health care, education, and recreation facilities. IETs will also supervise service employees. Typical job classifications are *inspector, production controller,* and *warehouse manager.* Health and regulatory inspectors will be needed to enforce a wide range of regulations concerning service industries.

In whatever capacity they serve, IETs work mainly with people. To be successful they must enjoy dealing with others. They must also understand the overall organization and priorities within their business.

The *Economic Research Institute* reported that the average starting salary was $32,000 in 1995 for an industrial engineering technician. The average salary was $36,500 for all workers, including new workers in this field. Experienced technicians earned $41,500.

Personal qualities needed in order to achieve as an IET include

- a willingness to work as part of a team
- an aptitude for mathematics including statistics
- an ability to do detailed work with a high degree of accuracy
- an ability to manage effectively with good verbal skills

Manufacturing Engineering Technician

In many technical colleges the industrial and mechanical (see following section) engineering technologies are blended into a *Manufacturing Engineering Technology* program. Demand for manufacturing engineering technicians and technologists should be favorable, even though the percentage of the U.S. workforce involved in manufacturing has been slowly declining. Increasing manufacturing engineering programs will provide graduates to assist in making future manufacturing activities more productive and more competitive in global markets. The manufacturing engineering technician should have the following:

- a strong work ethic and a good knowledge of quality principles,
- good report writing skills coupled with competency in oral communications (Figure 2.13),
- the ability to form, lead, and work well with teams,
- a strong understanding of computers and of computer-integrated manufacturing,

Figure 2.13 The manufacturing engineering technologist explains a manufacturing process. (Courtesy of Metatec® Corporation)

- a variety of basic IET skills and knowledge, such as costing and pricing, material selection, and safety,
- a variety of mechanical engineering technology skills and knowledge, such as basic turning, milling, grinding, extrusion, and welding operations.

2.6 Mechanical Engineering Technician

The *mechanical engineering technician (MET)* works with the muscle of industry—power transmission equipment. Some examples of such equipment are lever systems and gear trains, hydraulic and pneumatic systems, pumps and compressors, lathes, and milling machines. The MET calculates forces such as stress or strain on machine components, sets tolerances on shafts and bearings, and measures vibration. Often, the MET must specify equipment by the type of tooling necessary for the job and by researching the cost of comparable models.

Six excellent career fields within mechanical engineering technology are heating, ventilating, and air conditioning (HVAC); mechanical design (tool design); numerical control (NC); technical sales; fluid power; and laser technology.

Heating, ventilating, and air conditioning technicians aid in designing, manufacturing, specifying, installing, and maintaining HVAC systems (Figure 2.14). HVAC

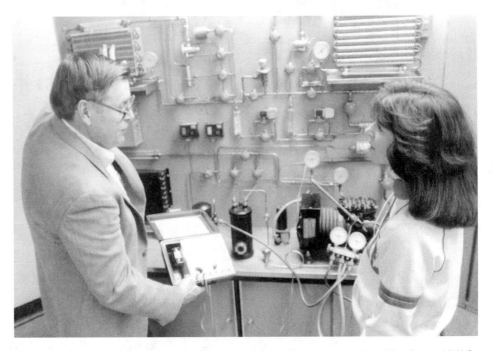

Figure 2.14 An HVAC instructor demonstrates the refrigeration trainer. The future HVAC technician will work with it to understand how a typical refrigeration system works. (Courtesy Central Ohio Technical College)

technicians must work with a wide range of skilled workers and engineers. In the manufacturing area, HVAC technologists are involved with product development and testing. They also contribute to the production areas where the equipment is built. Many manufacturing processes in HVAC are becoming automated, and those interested in careers in this segment of industry will need to understand computer-assisted production.

In the contracting and service area, HVAC systems must be properly mounted on solid foundations, they must be properly wired and plumbed, and they must meet the needs (specifications) of the building or facilities they are designed to support. Construction drawings must detail the construction plan. ASHRAE, the American Society of Heating, Refrigerating, and Air-Conditioning Engineers, Inc., lists the following job titles that technicians and technologists may qualify for:

Outside sales	Job foreman
HVAC inspector	Application technician or technologist
Service manager	Field service technician or technologist
Estimator	Energy conservation technologist

Mechanical design technicians are the communicators of industry. They prepare—or supervise the preparation of—assembly drawings that show how machine parts (e.g., jigs, fixtures, dies, cutting tools) fit together. A separate drawing is made of each tool (tool design) and of the completed machinery. These drawings are detailed and show several views of the machinery or part. Correct dimensions with tolerances are included. (See also Section 2.2, Drafting and Design Technician.)

The *mechanical design technologist* often supervises skilled drafters, referred to as detailers and checkers. A *senior design technician* is responsible for (1) checking assembly drawings for accuracy and completeness, (2) specifying materials on the basis of strength calculations, and (3) other considerations such as cost and availability of materials and purchased parts.

The industrial design area uses CAD extensively. With a CAD system such as the one shown in Figure 2.15, the technician can sketch ideas on a CRT. Once the drawing is satisfactorily displayed, a hard-copy printout of the drawing can be made by the computer's printer.

Many times the MET is responsible for connecting (interfacing) a computer to the machine process. Computer/machine interfaces are referred to as numerically controlled systems. *Numerical control* (NC) machining makes automation possible. NC machinery has two basic components: (1) an electronic controller (microprocessor) and (2) a machine tool.

NC technician openings will increase much faster than the average for all occupations through the 1990s. The *NC technician* must be somewhat familiar with the world of electronics as well as being on speaking terms with several types of computer languages and specific computer controllers (Figure 2.16).

The CAD system may also send instructions immediately to NC machines that can manufacture the designed part. This latter step, already reality in high-technology industries, is referred to as *computer-aided manufacturing* (CAM). A completed marriage of CAD and CAM results in a CAD/CAM system, or CADAM system.

Figure 2.15 A CAD system quickly changes color patterns to assist textile manufacturers to create new weave patterns. (Courtesy International Business Machines Corporation. Unauthorized use not permitted.)

When many of a plant's computer-controlled workstations are interconnected, a *computer-integrated manufacturing* facility results. This remarkable system is known as CIM (pronounced sim), and full implementation will constitute a challenge for technologists in the near future.

Technical sales representatives market their company's products to other manufacturers, wholesale and retail establishments, and government institutions. Companies supplying machine tools train MET graduates in the specific attributes of their products. The informed technician then calls on industrial clients (purchasing agents) to demonstrate how the products can increase or improve the customer's productivity.

Technical salespeople usually have large territories and must travel frequently. They work long hours, but may plan and work their own schedules. There is pressure on them to sell because they almost always work on commission. However, sales commissions can often exceed the salaries of other technologists. The best opportunities in this field will be for good communicators and for individuals with the desire to sell. Employment growth will be at least as fast as the average.

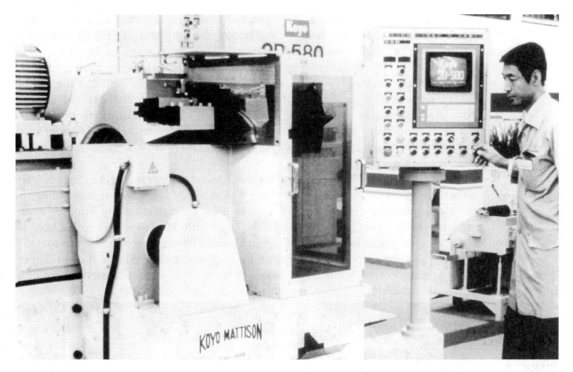

Figure 2.16 This automated grinding machine is programmed by an NC technician. (Courtesy G. E. Superabrasives)

EXAMPLE 2.4 Research a specific occupation within the broader occupational area of Mechanical Engineering Technology (MET) that is projected to have good job opportunities over the next decade.

Solution According to the *Occupational Outlook Quarterly,* Spring 1994, page 42, the specific occupation of *Tool Programmer* will find good job opportunities because, in recent years, employers have reported difficulties in attracting workers to machining and tool programming occupations. Tool programmers should earn at least $500 per week. The same journal, on page 38, reports an increase in employment "faster than average" for heating, air conditioning, and refrigeration technicians. Well-educated and competent HVAC technicians enjoy good employee benefits and earn as much as $700 per week.

A career in *fluid power* offers unparalleled opportunities for advancement and a good salary. Fluid-power systems use air or oil, are inherently dirty, and become untouchable to many. What may be an unglamorous occupation to most of us can be very profitable for those few technicians willing to put up with the inconvenience of working around unclean machinery. Fluid-power technicians often design systems. The mathematics of simple fluid-power systems need not involve more than algebra and trigonometry, and extensive tables may be found in handbooks to assist in design. The fluid-power technician also enjoys positions in technical sales and troubleshooting/maintenance.

The National Fluid Power Association (NFPA) reported in 1988 that the United States was the leading producer of fluid-power products, with a 46 percent share of the total world market. Next in line was Japan with 19 percent of the world market. The NFPA report also stated that the $6.7-billion industry in the U.S. is larger than many better known industries such as mining machinery, robots, tractors, and metalworking machinery.

The latest, most promising, field for mechanical engineering technicians is *laser technology*. Everyone has witnessed the use of the laser as an optical mark reader to check out groceries (Figure 2.17A). This is done by reflecting the intense single-frequency light waves generated from the laser off code strips on the package to photodetectors under the counter. Laser light is also used in communications to transmit voice, picture, and other information through fiber-optic systems. It is used in industry to weld, cut parts (Figure 2.17B), or connect small wires. It is used in medicine

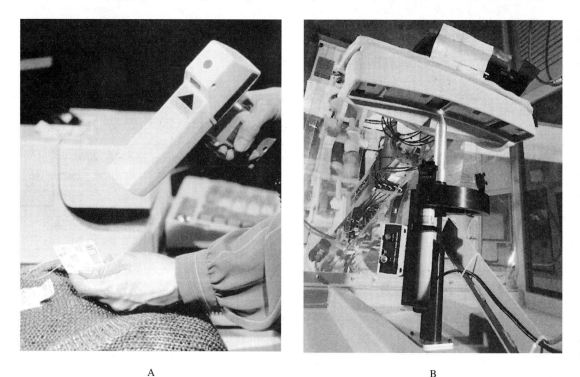

A   B

Figure 2.17   (A) The laser is used to read product information from bar-coded strips. (Courtesy International Business Machines Corporation. Unauthorized use not permitted.) (B) The helium-neon laser cuts the automobile part while the robot holds the part. (Courtesy Cincinnati Milacron)

(biomedical technology) to perform delicate surgical operations. The laser is used also for military purposes (e.g., the "Star Wars" missile defense program).

The *Economic Research Institute* reported that the average starting salary was $28,500 in 1995 for a mechanical engineering technician. The average salary was $35,000 for all workers, including new workers. Experienced workers earned $40,500.

Personal qualities needed in order to achieve as an MET include

- a willingness to work with others as part of a team and to direct others
- an aptitude for mathematics and science
- an ability to work with care and accuracy
- creativity in design work
- mechanical aptitude

Problems

1. Write out from memory what the following abbreviations of the major technical areas stand for: ChET, CET, EET, CpET, IET, MET. Expect your instructor to test you on this item.

For problems 2 through 8, research, and describe in your own words, the nature of the work for the following areas. For your research, try a career information delivery (CID) system such as *Discover* (see Section 1.3). CIDs may be found in your college library or placement office. Take a position, with listed reasons, on whether or not you would enjoy working in that area. Check with your instructor as to whether to prepare a one-page report or to submit a computer printout.

2. Report on one of the eight areas of specialization in Section 2.1, Chemical Engineering Technology.

3. Report on either surveying or drafting and design occupations (Section 2.2).

4. Report on the position of biomedical equipment or laser and fiber optics technicians in the electronics or electromechanical areas (Example 2.2).

5. Report on the duties of the data communication technician (Section 2.4).

6. Report on the position of inspector in the manufacturing sector (Section 2.5).

7. Report on the position of manufacturing engineering technician (Section 2.6).

8. Report on one of the eight areas of specialization in Section 2.6, Mechanical Engineering Technology.

9. Research and describe one specific occupation in the engineering technologies that is expected to expand within the next few years (see Example 2.1).

10. Research and describe one specific service industry occupation in the technologies described in this chapter that is expected to expand within the next few years (see Example 2.3).

11. List from memory the major engineering technology areas, and their abbreviations, offered at your community or technical college. Which of these rely heavily on electronics? Expect your instructor to test you on this item.

12. Make an appointment with an instructor at your college who is instructing in a technology of interest to you. Ask about the curriculum in that program and request to see the laboratory. Submit the curriculum for that program to your course instructor.

For problems 13 through 29, classify each job function as best performed by the ChET, CET, EET, CpET, IET, or MET.

Job Function	*Classification*
13. performs NC programming	_____
14. maintains large computer system	_____
15. supervises composite materials batch mixing	_____
16. performs time study	_____
17. troubleshoots and repairs mainframe computers	_____
18. designs a part for a fishing reel on a CAD system	_____
19. specifies HVAC system for model home	_____
20. constructs SPC chart to monitor a process	_____
21. maintains and operates a radio station	_____
22. prepares final assembly drawings	_____
23. explains machine tool to potential customer	_____
24. surveys the route for a new highway	_____
25. inspects buildings for code violations	_____
26. assists in testing new drug for harmful side-effects	_____
27. installs a fluid-power system	_____
28. maintains a microwave system	_____
29. inspects wastewater treatment plants for code violations	_____

30. Interview someone working in a technical area of interest to you. Write a one-page description of his or her typical workday.

Selected Readings

Curriculum materials describing the programs offered at your college.

Career Information Center, 6th Edition, Volume 6, 1996. New York, NY: Simon and Schuster: 1996.

3

Survival Skills—Preparing for the Engineering Technologies

John came to the technical college the August following his graduation from high school. He was planning to attend a four-year engineering college, but somehow never got around to applying in time. Perhaps his inaction was due to his not being sure he wished to embark on a long, arduous, theoretical education. John was not lazy; he was always tinkering with something and liked to understand how equipment worked, but he was most interested in hands-on applications, not simply theory.

His father, an engineering technician in glass manufacturing, was disappointed, but realized that John was practically oriented like himself. John's father learned about the local technical college's engineering programs and suggested that his son apply there. John attended an orientation that summer after high school, found out how practically oriented the engineering technology programs were, and decided to attempt the shorter, two-year curriculum. He decided to major in electronics engineering technology. John was successful and graduated with honors two years later. He is now employed as an instrumentation technician (Figure 3.1).

Pat graduated ten years before she came to the technical college. Recently divorced and a single parent, she hoped to improve her rusty mathematics and science skills to find a higher-paying job. The only jobs she could find before coming to college were minimum-wage jobs that could never pay for the kinds of things she wanted for herself and her five-year-old daughter.

Her grants came through while she was in summer school, and she found the mathematics refresher course easy. Pat completed her two-year technical degree in electronics and was employed as an electronic technician. She is presently pursuing a four-year technology degree. She feels the return to college has been a real stress-buster. The classroom gives her an opportunity to discuss her work and explore creative ways to improve herself. Pat feels the technical college opened doors for her that would have never opened without the practical education she gained at the technical college.

Louis came to the technical college after being laid off from his position as a mechanical engineering technician. His references from his supervisor were excellent.

Figure 3.1 After graduation, Louis was employed as an instrumentation technician. (Courtesy ISA, The International Society for Measurement and Control)

Louis was described as a good technician who could be depended on. However, Louis had never attended college. His high school education seemed enough, at the time he began working, to allow him to keep up with his previous employer's technology. Now Louis had to find a new job to support his wife and two children, and he was competing with technical college graduates.

The local technical college not only accepted Louis, but showed him how to apply for grants and scholarships. Louis also learned that he could pass proficiency tests in many of the beginning mathematics and pre-technical courses because he had had to use this knowledge in his previous job. Less than two years later he graduated with honors and had a better job than his previous one within a month of graduation.

These examples represent the broad diversity of backgrounds and experiences of today's technical college students. You may be able to identify with one of the examples above, or you may have a different set of experiences. Whether you recently graduated from high school, are experiencing a career change, or simply need to do something to change your present lifestyle, the engineering technologies can be the solution you have been searching for.

Because you are reading this text, you probably have already decided to attempt a career in the engineering technologies. In the last chapter you discovered how important it is to be sure you are on the right track. This chapter addresses how you can be successful in studies that will prepare you as an engineering technician or technologist.

3.1 Pursuing the Technology Degree

The college-trained engineering technician or technologist is a relatively new phenomenon, created by the rapid technological advances in the 1960s and 1970s. Most technical colleges—both public and private—existing today did not exist in the 1950s. The technician before 1960 was schooled in the military, trained in apprenticeship programs, or simply provided with on-the-job training (OJT). These informal and less theory-intensive educational programs did not provide experiences necessary to understand the many new materials and processes being developed. Such programs did not provide the intensive course work in communications and math-sciences that is vital to success and advancement in today's increasingly complex business and industrial world.

Modern technical colleges provide broad-based curriculums that better prepare technicians and technologists for the challenges of a more sophisticated world. There are two important degrees offered by these colleges:

1. The two-year degree is the **Associate of Science in Engineering Technology** (A.S.E.T.) degree (Figure 3.2A), and the graduate carries the position title of technician.
2. The four-year degree is the **Bachelor of Science in Engineering Technology** (B.S.E.T.) degree, and the graduate is referred to as a technologist.

The master of science in engineering technology (M.S.E.T.) is not considered here, but an increasing number of such programs exist. Two-hundred and ninety three persons were enrolled full-time in M.S.E.T. programs in 1989. Many of these graduates will become future technical college faculty.

The central purpose of the A.S.E.T. and B.S.E.T. is to prepare students for jobs. Both programs offer experiences with industrial-grade equipment and machinery. Hands-on operation of equipment is vital. Without real-world experience on current and up-to-date equipment, the technician cannot be productive in the short time period industrial managers expect (usually 30 to 60 days).

Most engineering technology students enroll in a two-year associate degree program. For many, though, this will be only a step towards a four-year program. Most

Figure 3.2   (A) The Associate of Science in Engineering Technology (ASET) is earned after two years of full-time study.

two-year colleges have transfer agreements with colleges offering the B.S.E.T. It is probably wise to check into such agreements early in your education, so you may best plan your future. It is more important, however, to make sure you can reach your present goals. The two-year degree is more easily attainable, and will result in many career opportunities, but the four-year degree will command higher salaries. Please refer to Example 1.2 for salary information.

Students who wish to attend a traditional college or university will enroll in a Baccalaureate of Science (B.S.) degree program rather than enrolling in A.S.E.T. or B.S.E.T. programs. One major difference in the B.S. degree track is that the first two

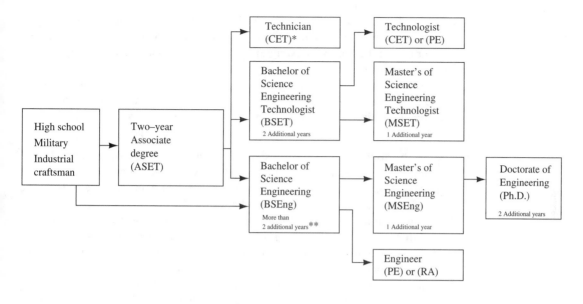

* See certification in Section 3.4.
** For ASET graduates.

Figure 3.2 (B) Career education pathways.

years are spent fulfilling general academic requirements. This difference usually results in the student taking more than an additional two years to earn the B.S. degree.

The first two years of the A.S.E.T./B.S.E.T. blends laboratory work with general academic courses in the student's selected technology. People interested in continuing their education towards the Master of Science (M.S.) degree should investigate opportunities at institutions offering the B.S. degree. Figure 3.2B illustrates the different educational tracks you might consider in order to fulfill individual goals.

The Curriculum

The major engineering technology classifications listed in Chapter 2 represent only a few of the more specific curriculums offered by your community or technical college. These colleges will offer a large variety of programs geared to local needs. For instance, technical colleges in Maryland offer such programs as aviation maintenance, chemical technology, construction technology, and electrodiagnostic technology—all of which are well represented in the state's industrial base.

No matter how specific, all A.S.E.T. programs will require a core curriculum of *general studies* courses—for instance, communications, psychology, and sociology. The difference between general studies in the two-year versus the four-year colleges is that the four-year liberal arts curriculum has more courses in general studies. A.S.E.T. programs must focus their much fewer credit hours on courses that teach students how to accomplish tasks essential to their fields. The few general studies courses you will be exposed to will be very important to your ability to advance in your career. Item number one in the following paragraph should serve to illustrate the importance of general studies to your career.

A national dialogue concerning technical colleges (*National Roundtable on Economic Development,* July 1987) describes the characteristics employers are looking for when they hire technicians. These hundreds of employers interviewed required technicians who

1. have the ability to read, write, listen, speak, and work with others in technical teams, with others in the organization, and with customers;
2. have a strong base in applied math and science and are capable of learning new specialties as the technology changes;
3. are adept in the use of computers for data acquisition, storage, manipulation, and display; for automated control of machines; and for use in design;
4. possess a combination of knowledge/skills in mechanical, electrical, fluid, thermal, optical, and microprocessing devices; and
5. understand how electronic and mechanical systems and subsystems are interrelated. Your A.S.E.T. curriculum will be based on these or similar requirements.

Cooperative Education

Pursuing a degree in the engineering technologies can be made easier by enrolling in a cooperative education (co-op) program. Early in the twentieth century, technical education began with the offering of co-op programs at the University of Cincinnati and Northeastern University in Boston. These institutions formed college-industry programs that would keep the best of theory and practice, by having students alternate semesters in college with periods of work in industry. Today, many of our technical colleges have similar programs. Your instructor will be able to inform you of any co-op programs at your college.

To enter a co-op program, you must be accepted by the employer. This means that you will have to prepare for one or more interviews (Figure 3.3). Some typical questions an interviewer might ask are

- Why should we hire you?
- What is your greatest strength? Weakness?
- Where would you like to be in two years? Five years?
- What led to your interest in this career area?

The best way to prepare for an interview is to have formed a realistic opinion of yourself. Self-assessment is difficult for many people, especially young students just out of high school. One of the ways to best capture who you are—your skills, interests, and enthusiasm for your chosen career—is to build a résumé. You must have a résumé to qualify for most co-op programs.

The good news about building your personal résumé is that it must be short. Studies show that human resources personnel, the people who usually first evalu-

Figure 3.3   A co-op student is interviewed. (Courtesy International Business Machines Corporation. Unauthorized use not permitted.)

ate résumés, spend fewer than two minutes screening résumés for advertised positions. It is recommended that most potential employees keep their résumé to a single page. Of course, a brief résumé must be well thought out, interesting, and attractive to the reader. Check out the following suggested outline, with comments, for a résumé:

OBJECTIVE   A brief statement that will get the employer's attention. Example: To be employed as an electronic engineering technician.

EDUCATION   Example: Candidate for Associate of Applied Science Degree in Electronic Engineering Technology. Graduated in upper 30 percent of my high school class, while working part-time at a local electrical/electronics warehouse. Elected class president in my senior year.

BUSINESS
EXPERIENCE   1994–1996 Worked 16 hours per week as an inside salesperson. Interacted directly with customers. Received two raises, one in my junior year of high school and the second in my senior year.

PERSONAL   Born in 1980, and raised in North Carolina. In excellent health, 5′8″ tall, 155 pounds, single. Travels include the Western United States and Canada. Hobbies are fishing and backpacking.

REFERENCES   Personal references available upon request.

Finally, remember to list and highlight your past successes. If you served in any organizations as an officer, that shows the employer that you can get along with people. It is often difficult to understand that you probably already have a wealth of experience that will make you a great employee. Search hard for those experiences and accomplishments that you take pride in and make you who you are.

Computer software is available to aid you in formatting your résumé. Check with your career services or placement office.

The on-campus interview will be a screening interview. In order to prepare for the interview, you should think of how you wish to present yourself. You will of course wish to dress professionally and be well groomed. Presenting a good first impression is vital to your success. You should also have some knowledge of the company you will interview with. What is their product? Where are they located? How long have they been in business? You should also be prepared to ask questions. What skills and knowledge will be expected of you? Will your work periods fit with your college course obligations? How will your work be evaluated? What will be your salary and benefits? What training and development programs are available to co-ops?

An on-sight interview at the business or industry's location may be required. Expect to meet people that you will be working with. It is essential that you strive to remain calm and poised, reflecting an ability to work well with others and especially in problem-solving teams. After each interview, be sure to ask for the interviewer's business card. Follow up with a polite thank you letter.

As you prepare for a co-op position, your co-op coordinator or technical instructor will be your best resource. They will help you construct a résumé, and ready you for the interview process.

Planning for Graduation

Plan for graduation when just beginning your degree program? Ask your career services or placement professional. Technical and community college placement officers say that it is never too early to begin planning for your job search. Most will ask you to visit the placement office early, in order to know what information you will need to have upon graduation. You will also find out what career information services they offer, if any software is available to create résumés, information on local businesses and industries, local job fairs you should attend, and part-time positions available. You will also find that <u>you must register with the placement office at least a quarter or semester before graduating.</u>

An excellent idea is to begin immediately to assemble a *graduation portfolio.* A large manila folder can be used to organize the material to be included. The portfolio will contain items you will wish to refer to later as you prepare your résumé and your job search strategy. Suggested items include

- course syllabi, in order to refer to the performance objectives you have mastered.
- project reports, especially those that were prepared with other students in a team situation.
- a continuously updated résumé.

- a list of companies you might want to work for along with contact persons in those companies and annual reports (often obtained free from local investment firms).
- personal notes, your thoughts over time regarding your career aspirations.

You may have already surmised from the above comments that a part-time job, especially if it is in a technical field, will offer extremely beneficial résumé material. Even a part-time job in food service or as a clerk in a discount store can demonstrate desired personal characteristics such as dependability, punctuality, and self-motivation.

Succeeding in College Life

The challenge for technical educators is to supply you, the aspiring technician, with the skills and knowledge it will take to have the characteristics employers need.

The challenge for you is to acquire them. How do you cope with the enormous challenges of college life—especially in the first few months, before good study habits and other campuswise skills are developed? There are two important prerequisites for success:

1. maintaining a positive attitude and
2. planning how to meet day-by-day and week-by-week goals.

Maintaining a positive attitude is based on an intrinsic motivation to succeed in all of your endeavors. This intrinsic or internal motivation must be developed, and this is most easily accomplished by associating with those who have like interests and goals. Some colleges have professional organizations or groups that aid students in forming such relationships. The importance of professional organizations in achieving your career objectives is discussed at the end of the chapter.

The second prerequisite to success is *planning*. Listening carefully to the instructor in class is an important step in planning. The instructor informs students of deadlines for homework and dates of quizzes and exams. Careful listening and note-taking assure you of meeting the basic goals of the course. Planning also involves scheduling study periods for each class (Figure 3.4). A good rule of thumb is to schedule approximately two hours of study for each hour of lecture.

Planning can be facilitated by consulting your instructor, during regularly scheduled office hours, whenever questions arise that cannot be answered in class. Involving the technical instructor in your program and career cannot be overemphasized. Most students require help with planning and find it difficult to go it alone.

Communicating

Students seeking to become engineering technicians or technologists often lack communication skills. It has already been noted that employers regard good communication skills as a necessary element of technical positions. These managers know that a paraprofessional on the engineering team must be able to communicate ideas to others. This is why engineering technology instructors require good report writing in technical courses. In Chapter 7, laboratory report writing is addressed, but it must be stressed here that good report writing includes no spelling errors. This point is so crucial that a list of the most commonly misspelled words in engineering reports is offered in Table 3.1.

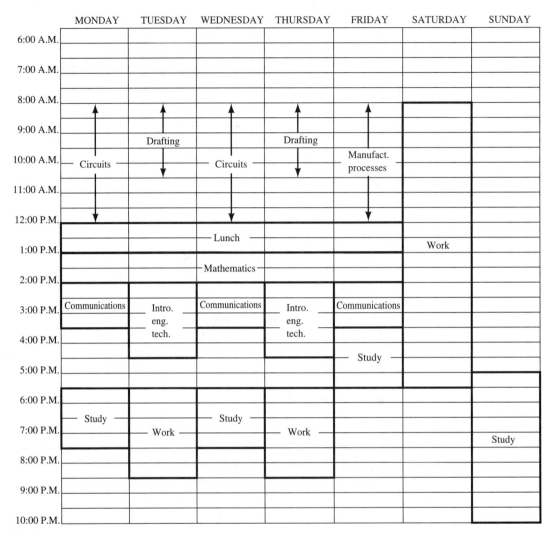

Figure 3.4 A typical technical student's schedule is laid out to assure adequate study time.

Good spelling, as well as good writing, takes discipline, dedication, and attention to detail. The conscientious student consults the dictionary or other resources whenever the spelling or usage of a word is in doubt. A report that has excellent content will not be seriously considered by most readers if they see misspelled or misused words in the report, especially if the words are technical in nature.

Mathematics and Science

The *math-sciences* form a critical base for the technician and technologist (Figure 3.5). You must be able to work intelligently with industrial machinery, materials,

Table 3.1 The Most Commonly Misspelled Words in Technical Reports

1. recommend	11. hydraulic	21. discrepancy
2. vacuum	12. pneumatic	22. heat-treat
3. precede	13. proceed	23. humidity
4. basically	14. separate	24. materials
5. volume	15. inaccurate	25. procedure
6. proportional	16. maintenance	26. necessary
7. comparison	17. auxiliary	27. comparative
8. occurrence	18. develop	28. process
9. nozzle	19. mechanical	29. electromechanical
10. resonance	20. reservoir	30. receive

Figure 3.5 The technician must know the math-sciences to work effectively. In this case, a technician uses a calculator on a construction project. (Courtesy Hewlett-Packard)

and processes. A good grasp of physics, chemistry, and the supporting mathematics is the only way to understand complex and highly theoretical systems. The applied math-science courses you will encounter are not watered down. The term *applied* merely means you will not encounter as many proofs and that relevant engineering examples and/or laboratory experiences will be integrated with the theoretical material.

Attention to detail is emphasized in the example of the spelling list in Table 3.1. In learning math-science principles, attention to detail is essential also. *The only way to understand the rigorous principles of the math-sciences is to solve problems!* This text offers many worked-out examples. You can begin to appreciate the need to do many problems by attempting to solve the text's example problems without reading the author's solutions. After you have attempted a problem or arrived at a solution, consult the author's solution.

Faithfully completing your homework in a regularly scheduled study session soon after each lecture will also lead to success in learning the principles of the math-sciences.

3.2　Using the Engineering Library

A strong base in the math-sciences, coupled with good communication skills, allows you to make efficient use of the library. The technical college library contains general and technical reference works. General reference works include dictionaries and encyclopedias, which you are already familiar with. You know they are most useful as starting points for exploring areas that you have little knowledge of.

A general reference work that you may not be familiar with is a government publication, the *Dictionary of Occupational Titles,* often referred to as the DOT. The DOT is an excellent career reference. It contains a listing of standardized occupational titles and the responsibilities of the positions. Figure 3.6 is a section from the DOT. This section describes the rather broad position of Test Technician.

The value of the DOT is its use of the language of personnel officers. This is the language that you should use in your résumé. *College placement counselors recommend that you begin to prepare a résumé as soon as you begin your technical education.* As your course work progresses, you will know how the various technical skills and knowledge you acquire can best be worked into a complete résumé. When you graduate, your résumé will be ready to distribute to the employers you are interested in.

Technical skills can also be improved in the library. The technical college library contains books, handbooks, and periodicals (magazines) explaining practical skills as well as more theoretical topics. The technician or technologist who ignores library resources is at a distinct disadvantage when competing in the real world of industry.

No matter what your field of specialization, there will almost surely be a handbook devoted to that field. You'll find such titles as *ASHREA Handbook, Fundamentals Handbook of Electrical and Computer Engineering, Handbook of Air Pollution Technology, Handbook of Industrial Robotics, Instrument Engineers*

019.161-014 TEST TECHNICIAN (profess. & kin.)

Prepares specifications for fabrication, assembly, and installation of apparatus and control instrumentation used to test experimental or prototype mechanical, electrical, electro-mechanical, hydromechanical, or structural products, and conducts tests and records results, utilizing engineering principles and test technology: Confers with engineering personnel to resolve fabrication problems relating to specifications, and to review test plans, such as types and cycles of tests, conditions under which tests are to be conducted, and duration of tests. Fabricates precision parts for test apparatus, using metal working machines, such as lathes, milling machines, and welding equipment, or interprets specifications for workers fabricating parts. Examines parts for conformance with dimensional specifications, using precision measuring instruments. Coordinates and participates in installing unit or system to be tested in test fixtures, connecting valves, pumps, hydraulic, mechanical or electrical controls, cabling, tubing, power source, and indicating instruments. Activates controls to apply electrical, hydraulic, pneumatic, or mechanical power and subject test item to successive steps in test cycle. Monitors controls and instruments and records test data for engineer's use. May recommend changes in test methods or equipment for engineering review. Workers are classified according to engineering specialty or type of product tested.

Figure 3.6 Job description from the *Dictionary of Occupational Titles* (DOT).

Handbook, Machinery's Handbook, 25th ed., Standard Handbook for Civil Engineers, 3d ed., and *Welding Handbook, 7th ed.*

The most recent technological advances are found in magazines (often referred to as *journals* or periodicals) in your college library. These magazines will include titles, such as *Electronics, Plant Management, Foundry, Hydraulics and Pneumatics, InTech,* and *Mechanical Engineering.*

These magazines are full of relevant and current information. Figure 3.7 shows how to access articles that are listed in the *Applied Science and Technology Index* (an index specifically targeted to technical magazines), by using your college's library computer system. Figure 3.7A begins with the *Subject Referenced Databases* in this particular computer system. The *Applied Science and Technology* database is selected on this screen. The next screen (Figure 3.7B) asks how you will search the index, and *Subject* is the selection. In this case, the subject screen was completed by typing in *Robotics,* and a list of titles is presented in Figure 3.7C. The title *Beyond R2D2* is selected resulting in the screen (Figure 3.7D) showing an augmented (expanded) title and other pertinent information. At this time the user would, if still interested, search for the November 1996 *Byte* magazine or consult with a librarian on how to obtain the article from other libraries. Most libraries will obtain faxed materials from other libraries for a small fee. Remember that you cannot possibly read all of the technical magazine articles written. Focus on your interests by checking the *Applied Science and Technology Index* each month for articles within your area of interest.

Technical college librarians are eager to help students. In Figure 3.8, a resource librarian shows a technical student how to use a general reference work and find an engineering periodical. *Don't be accused of reinventing the wheel* or of spending time and money developing tools or processes already described in reference works and magazines. Use your college or business library.

Subject Reference Databases

	Interface
01 > ABI/INFORM (Business)	(III)
02 > AIDSline	(OVID)
03 > Anthropological Literature	(RLG)
04 > Applied Science & Technology	(III)
05 > Art Index	(III)
06 > Avery Index to Architecture	(RLG)
07 > BioethicsLine	(OVID)
08 > Biological Abstracts	(OVID)
09 > Biological & Agricultural Index	(III)
10 > CancerLIT	(OVID)
11 > CINAHL (Nursing)	(OVID)
12 > Compendex	(OVID)
13 > Education Index	(III)
14 > English short Title Catalogue	(RLG)

F > FORWARD
R > RETURN to previous menu

Choose one (1-14,F,R)

A

Welcome to Applied Science and Technology
Index
Covers 400 engineering, technology, and
science periodicals since 1983.
Last updated: 31 Dec 1996   845,582 records.
You may search the following:

I > INFORMATION about this database

W > WORDS in title/subject/abstract

A > Article AUTHOR
T > Article TITLE
P > PERIODICAL Title

S > SUBJECT
R > REPEAT Previous Search

X > EXIT from Applied Science and Technology
   Abstracts
Choose one (I,W,A,T,P,S,R,X)
For assistance ask library staff.

B

You searched for the SUBJECT: robotics

243 entries found, entries 1-8 are:

Robotics Research

		APPEARS IN	AST DATE
1	Beyond R2D2: robots evolve	Byte	Nov 96
2	Robots on all twos	Technology Review	Nov 96
3	Immobile robots: AI in the new millennium	AI Magazine	Oct 96
4	Autonomous vehicles	Proceedings of the I	Aug 96
5	A critical study of the applicability of r	Journal of Applied M	Jun 96
6	Active learning for vision-based robot gra	Machine Learning	May 96
7	Learning concepts from sensor data of a mo	Machine Learning	May 96
8	Learning controllers for industrial robots	Machine Learning	May 96

Please type the NUMBER of the item you want to see, OR
F > Go FORWARD   A > ANOTHER Search by SUBJECT
R > RETURN to Browsing   P > PRINT
N > NEW Search   + > ADDITIONAL options
Choose one (1-8,F,R,N,A,P,D,L,J,E,+)

C

Figure 3.7 Using the library computer system. (From *Applied Science and Technology Index.* Copyright © by the H. W. Wilson Company.)

You searched for the SUBJECT: robotics   AST
                                                                Record 1 of 243

TITLE   Beyond R2D2: robots evolve.
             Augmented title: interview with Takeo Kanade, director of the Robotics
             Institute at Carnegie Mellon University.

APPEARS IN   Byte v. 21 (Nov. '96) p48.

NOTE(S)   Interview.

SUBJECT(S)   Kanade, Takeo.
                  Robotics research

---

1 > Display HOLDINGS

Key NUMBER to see detailed holdings, OR
R > RETURN to Browsing   A > ANOTHER Search by SUBJECT
F > FORWARD browse   S > SHOW SIMILAR items
N > NEW Search   + > ADDITIONAL options
Choose one (1-1,R,F,N,A,S,P,T,E,+)

D

Figure 3.7   (continued)

A   B

Figure 3.8   The resource librarian helps a technical student to (A) use an encyclopedia,
(B) find an engineering periodical. (Courtesy Central Ohio Technical College)

3.3 Problem Solving

To be successful in college, the technical student must be a good problem solver. The dictionary defines a *problem* as "a *perplexing* question *demanding settlement,* especially when it is difficult or *uncertain* of solution." The key words are italicized. All problems demand settlement—require good problem solvers.

How often have you observed individuals refusing to make a decision or take action when decisive action is obviously needed? Is fate the only operative in this universe, or could better problem solving on the part of world leaders have changed history and improved the lives of all of us? How often have *you* failed to make a decision and act on that decision, causing yourself and those who depend on you to suffer?

There are many valid reasons that effective problem solving is so rare. Some of the most important reasons are

- a lack of information
- fear of making a wrong decision (or fear of failure)
- the time and energy needed to generate appropriate solutions and test the outcomes
- the absence of an effective problem-solving model

The first reason for not solving important problems, *lack of information,* can be overcome simply by realizing that we never have enough information. Problems become *perplexing,* or puzzling, because there is no neat solution that flows out of perfect knowledge. Good problem solvers enjoy the challenge of making decisions that others cannot or will not attempt. Being able to make these real-world decisions, made without full knowledge of the outcomes, is essential to your future happiness.

Fear of making a wrong decision, the second reason, may be easily put aside. The world's most successful people have failed. Winston Churchill, the great British statesman, was voted out of office after his decision to revaluate the British pound led to a general strike in 1929. He did not hold a cabinet position for 10 years after that, at times living with extreme depression. He went on to become the heroic prime minister of Great Britain during World War II. The most successful technologist will often make wrong decisions. If you feel you have never failed you have not attempted much and *actually have failed* to reach your potential. And remember that *failure can result in success if we learn from our failures.*

Your choice to work in technology means *you must find time and energy to solve problems.* You will be confronted with problems on a daily basis: with equipment that doesn't always work the way it should, supervisors pushing seemingly impossible deadlines, or irate customers who cannot understand why their business is suffering because of a faulty computer. Good problem solving does not require great intelligence or creativity as much as hard work. The most successful people in business and industry will tell you that problem solving is "99 percent perspiration and 1 percent inspiration."

A top executive officer in the telecommunications industry discussed the necessity for hard work to overcome shortcomings:

> To be honest, I like some days more than others, and some days I truthfully believe that the company cannot possibly survive my rash of bad judgments, but you will never find me quitting. Why? Because that is *my strength.* That is what I bring to the table. I have failings. I make mistakes. But no one, not ever, will outwork me. No one, not ever, will stick to a task longer than myself. No one, not ever, will out-plan me, will out-detail me, or outperform me. Again why? Because those are my strengths! Am I an intellectual giant? Not hardly. Am I an outstanding design engineer? Not in your or my lifetime. Is my English always correct? Don't I wish. But am I street smart, practical, and *hard working?* You bet!
>
> Jack A. Shaw
> Capitol College News
> Graduation 1994

The final reason, *absence of an effective problem solving model,* is simply not true. Behavioral sociologists have come up with models that have been proven to be effective. Consider the following basic elements of problem solving:

1. Recognizing there is a problem
2. Defining the problem
3. Brainstorming to determine alternative solutions (often a research step)
4. Considering the possible consequences of each solution
5. Selecting the best solution
6. Implementing the solution
7. Monitoring and evaluating the action taken

The need for early planning for graduation was discussed in Section 3.1. Put yourself in the place of an entering first-year Drafting and Design student who knows she must stay in the local area after graduation because of family commitments. Using the basic problem-solving elements, she reasons as follows:

1. The problem is recognized and defined—she needs to find local employment after graduation.
2. In brainstorming alternative solutions she comes up with the following alternatives:
   - do nothing until graduation and hope to find a local position, or
   - begin immediately to determine her chances for local employment in her chosen technology and how to successfully find a position.
3. After considering the consequences of each alternative, the student knows that the second alternative is the only logical choice.
4. In implementing the solution she consults with her technical instructors and the placement office to identify local companies that are potential employers. The placement office helps her to contact key personnel at the companies. One of the companies not only gives her assurances of their need for her after graduation, but offers her an immediate part-time job in her technology.

The job will not conflict with her class schedules, allowing her to continue her full-time education.

5. She monitors the decision to take the part-time job and concludes after two semesters that, on balance, the work experience helps her more than the time she must spend at work and away from her studies detracts from her success in the classroom. Her chances of gaining local, full-time employment after graduation are definitely enhanced.

This simple example is followed by Examples 3.1 and 3.2. These examples are more complex and may better illustrate the power of the seven basic steps in problem solving.

EXAMPLE 3.1

Sharon is a first-quarter technical student. She has completed three laboratory assignments and submitted the reports to the instructor. The instructor has returned the reports and each report has a grade of D marked at the top. No other marks or explanations are to be found on the reports.

Solution

STEP 1: Recognize the problem.

Sharon must first decide that she has a problem. She could pass off the Ds as unimportant, especially if she is receiving higher grades on written examinations. However, she is being trained in an occupation that requires hands-on experience and quickly realizes she does have a problem that must be confronted. Sharon decides to not turn in the next laboratory report until she has attempted to solve the problem. This interim action must not take too long because the instructor demands prompt reports.

STEP 2: Define the problem.

Sharon decides to consult immediately with the instructor to determine exactly what her problem is. The instructor could have commented in writing on Sharon's reports, and Sharon would not have to ask what is wrong with the reports. However, this important feedback was not available and it is up to Sharon to find out why she is submitting less than satisfactory reports. Sharon maturely realizes she is ultimately responsible for the solution of her problems, and she must not wait until the instructor decides to "let her in on the secrets of good lab work and reporting."

Sharon meets the instructor during his regularly scheduled office hour and discovers that her reports are well written and have good supporting graphs and pictorials. Her only problem is with the measurement data she has reported. The data are incorrect and the instructor believes that Sharon has not even performed the assigned experiments. Sharon assures the instructor she has attended every lab and performed each experiment with her two lab partners. In the future she will make sure the busy instructor knows she is attending lab. Sharon has identified her problem as **reporting incorrect data** (she was merely recording the data the other lab partners were measuring), and she is now ready to proceed to the next problem-solving step.

STEP 3: Brainstorm.

Problem-solving experts agree that a crucial step in solving a problem is listing all the possible solutions to the problem. Some alternatives may be rapidly eliminated, but a good first step is to list them all. Sharon's brainstorming of alternative solutions to her problem results in the following list:

1. Drop the course.
2. Drop out of her technical degree program.
3. Report the seemingly impersonal and uncaring instructor to his department head.
4. Determine how she can collect correct data.

STEP 4: Consider the consequences of each alternative.

The first two alternatives are eliminated because Sharon cannot achieve her goal of becoming an engineering technician without an appropriate education, which includes this laboratory course. The third alternative solution depends not only on the college administrator hearing with a sympathetic ear but also in his taking corrective action. Even if this happens Sharon still may not achieve her main goal. Also, Sharon correctly decides that, even though the instructor may have some shortcomings when grading reports, he is definitely not incompetent. The instructor has an extremely sound knowledge of the technical area and demonstrates this expertise daily in the theory portion of the course.

STEP 5: Select the best solution.

Sharon selects the fourth alternative solution as the simplest and most direct in achieving her goal. She realizes that the two lab partners have taken all measurements while she has only recorded the data. In a sense she has been acting as secretary to the group and not taking full responsibility for the results (Figure 3.9). After further consideration, she concludes that she has not carefully organized the data during the laboratory, but collected it on scrap sheets of paper often unorganized and illegible. Also, she does not write the report until the weekend—long after she has collected the data.

STEP 6: Implement the solution.

Sharon pursues the following solution steps during the remainder of the course:

1. She consults with the instructor on the correct use of measurement equipment, the correct interpretation of meter scales, and the best ways to collect data.
2. She takes her own laboratory measurements.
3. She records her data neatly and confirms it with theory before leaving the laboratory.
4. She writes her report in the library, immediately following the laboratory while the lab procedure is fresh in her mind.

Figure 3.9 Sharon is not reading the meter but merely recording data. (Courtesy Central Ohio Technical College)

STEP 7: Evaluate the action taken.

By the end of the quarter Sharon feels much more comfortable in the laboratory and passes her lab practical exam with a grade of A. She has improved her laboratory report grades to a B average. She consciously evaluates her implementation step and realizes that with the available resources she had at the time, and realizing she began with a D average in the laboratory, she caused a positive change in a poorly understood situation. Sharon evaluates her implementation step as productive. In future lab courses she decides to improve on the steps she established in the present quarter.

EXAMPLE
3.2

The industrial engineering manager in an automotive assembly plant meets with the technical staff to determine whether to automate a welding operation.

Solution

STEP 1: Recognize the problem.

The manager, in recognizing the problem, reports to the staff that all costs must be reduced to remain competitive with increasing foreign competition. A particular welding operation attaches the front fenders of the automobile to the body. It is a repetitive operation that can be automated. Currently, two operators per eight-hour shift perform the welding operation.

STEP 2: Define the problem.

The staff defines the problem as:

show by calculation that a cost savings can be realized by automating the operation.

Also, the staff must decide how to automate the line, i.e., to use fixed automated equipment or to use a robot to perform the welding operation.

STEP 3: Brainstorm.

The staff determines the exact type of automation to be a robotic application. Fixed equipment that is automated to perform the same operation over and over again (*hard automation*) will not work. Two models of automobile are assembled on the same assembly line. Therefore, soft automation must be used. *Soft automation* involves a computer and robot that can be adapted or quickly reprogrammed to handle the different automobile models.

Further research and discussion results in the following three alternatives.

1. Do not automate and leave the welding to assembly-line employees.
2. Automate the operation and use one employee per shift to monitor the robotic system.
3. Automate the operation.

STEP 4: Consider the consequences of each alternative.

For industries to survive, costs must be minimized. Therefore, each of the alternatives must be evaluated in regard to cost.

The cost of alternative 1 is calculated as follows:

$$\$20/\text{operator-hour} \times 2 \text{ operators} = \$40/\text{hour}$$

Twenty dollars is a typical automobile worker's salary, including fringe benefits.

The cost of alternative 3 involves many factors, such as the initial price of the robot and controller, engineering modification costs, installation costs, personnel costs to install and maintain the robot (personnel training costs will be included in this figure), additional hardware costs, and end-of-arm tooling costs.

A robot installed above the assembly line may reach the weld areas on both sides of the automobile. This gantry robot (Figure 3.10) costs approximately $100,000. Installing and maintaining the robot system is calculated at 1.5 times the original cost of the robot.

$$\begin{aligned} \text{Installation and maintenance} &= 1.5 \times \$100,000 \\ &= \$150,000 \\ \text{Total cost of robot system} &= \$250,000 \end{aligned}$$

If the robot works three shifts per day, 365 days per year, for a five-year period, the cost of the robot over the five-year period (the useful life of most industrial equipment) becomes

$$\text{cost/hour} = \$250,000 \div (5 \text{ years} \times 365 \text{ days/year} \times 24 \text{ hours/day})$$
$$\text{cost/hour} = \$250,000 = \$5.71/\text{hour} \approx \$6.00/\text{hour}$$

A robot application in the automobile industry averages approximately $6/hour. Compare $6/hour with the $40/hour cost of using the operators! But productivity will improve as well. The robot, once installed correctly, will not

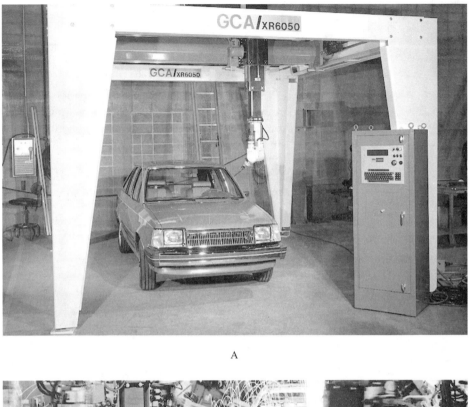

A

B

Figure 3.10 (A) A gantry robot can weld parts located over a large area. (B) Robots are used extensively in the auto industry to cut costs. (A, courtesy GCA Corporation, Industrial Systems Group; B, courtesy Cincinnati Milacron)

take coffee breaks, vacation, or sick leave. Also, a lower parts rejection rate adds to the benefits. On the basis of this example, is it any wonder that manufacturers choose to automate?

Step 5: Select the best solution.

Of course, the staff decides to automate. The decision, then, is whether to use alternative 2 (automate the operation and use one employee per shift to monitor the robotic system) or alternative 3 (simply automate the operation). Alternative 2 costs an additional $20/hour.

The industrial engineering manager, with the plant manager and the employee representative (union representative), note the jobs of six employees are in jeopardy (two operators per shift and three shifts per day). The labor/management team decides to continue with three of the original six employees to lessen the impact on the work force.

Step 6: Implement the solution.

The three employees travel to the robot manufacturer's plant and train to operate and monitor the robotic system. Simultaneously, the robot is purchased and installed. Set-up technicians ensure the capability of the system after installation. Preventive maintenance schedules are addressed.

Step 7: Evaluate the action taken.

After implementation the company found the following:

- The employees benefited by upgrading their skills and knowledge—greater pride of workmanship resulted.
- The plant has saved $14 per hour of operation. The savings for the year comes to

$$\$14/\text{hour} \times 120 \text{ hours/week} \times 52 \text{ weeks/year} = \$87,360/\text{year}$$

Example 3.2 reveals how industrial people work as a team in most problem-solving situations. The technician participates in the decision-making process with managers, engineers, and labor. The *team effort* (Section 1.2) requires the technician to be a good communicator and to respect the role others play in decision making.

Early Problem Solvers

An early electrical engineer, Charles Kettering (1876–1958), provides an example of the flexibility and excellent problem-solving ability of many engineers and technologists. He grew up on a farm, learning early how to use equipment and often how to improvise to get a job done. In college his weak eyes (myopia) forced him to study much more than most. Some of his friends read texts to him after classes.

Kettering's many accomplishments required skills and knowledge from many of the engineering fields listed in Section 1.1. He was responsible for more major

Figure 3.11   Charles Kettering works in his "barn workshop" to develop the self-starter that did away with dangerous hand cranks. (Courtesy Cadillac Motor Car)

inventions than any American other than Edison. One of his major inventions was the automobile self-starter (Figure 3.11).

The self-starter was necessary to eliminate injuries (and sometimes deaths) from crank starting. The 35-year-old electrical engineer naturally thought of an electric motor to do the job. Of course, he had to develop a generator and complete electrical system supported by battery to support the electric motor.

Kettering turned to a group of young, creative technicians known as the "barn gang" to implement his ideas. Timing was critical, for other inventors were working on the same problem. Also, Kettering and the barn gang could work only part time on the problem because of other work commitments at National Cash Register in Dayton, Ohio. The barn gang participated in the design, drafted the plans, and machined, wound, and wired the prototype. These creative technicians made the self-starter a reality.

One of the first female engineering graduates (B.S., Ohio State University, 1893), and the first listed in *Who's Who in Engineering,* was Bertha Lamme. Her work at Westinghouse in designing dynamos and motors earned her the reputation of being "a great problem solver." The *Pittsburgh Dispatch* reported in 1907 that, even in that hothouse of gifted electricians and inventors, "she is accounted a master of the slide rule and can untangle the most intricate problems in ohms and amperes as easily and quickly as any expert *man* in the shop."

Lamme had to respond to telephone calls from other engineers seeking technical advice in order to solve customers' problems. A sales engineer once called to report that he wanted to speak with her supervisor about a customer demanding major modifications to a motor. Lamme fielded his questions expertly, "Suddenly, he realized that the woman was redesigning the motor just about the way the customer wanted," stated Westinghouse historian Charles Ruch. "After he hung up the phone, the salesman ran over to the lab to see what kind of *secretary* could do that kind of thing."

SPC Problem Solving

Since the decade of the 1980s, many industries have used a problem solving plan developed by Dr. Walter Shewhart. Dr. Shewhart is known as the father of Statistical Process Control (SPC). The Shewhart Cycle is very similar to the general problem-solving steps developed in this section (see Figure 3.12A).

Industries depend on the Shewhart cycle and SPC to increase quality while decreasing costs. Take, for example, the use of the process-control chart, Shewhart's greatest invention. In Figure 3.12B, an automobile's gas mileage is plotted over time. As can be seen by the up-and-down nature of the points in the sequence, the gas mileage is variable; for instance, in-town driving produces poorer gas mileage than long trips on the interstate. But by viewing the variability statistically (each plotted point is an average of several gas mileage calculations), a definite downward trend can be seen. As the miles per gallon decrease there is a point, known as a control limit, where the cost of a tune-up is justified. (To spend money on a tune-up at, say, point 3 on the chart would be wasteful. The poor gas mileage in this case could be due to greater-than-normal driving in the city.)

Major auto manufacturers insist on their suppliers using SPC techniques. Figure 2.12 and the associated discussion serve as an example of the auto industry's dependence on SPC.

Attitude and Problem Solving

All of the formal elements of problem solving listed are worthless unless the problem solver begins the process with a good attitude. Some days, everything seems to go wrong and you cannot get along with anyone. Other days are great, nothing goes wrong, and everyone seems to agree with you. Most days, however, include a mix of these two extremes.

Why do you perceive your days to be so different? Some would argue that body chemistry, weather, biorhythms, or even the positions of the stars make the difference. Research, sometimes, appears to bear these arguments out. But many others contend that attitude during the course of a given day may be controlled by your own thoughts.

The *effective problem solver develops a positive attitude.* A positive attitude simply means saying "yes" instead of "no" to questions concerning one's ability. You must learn to attack each problem with the response, "I will find a way to solve this problem."

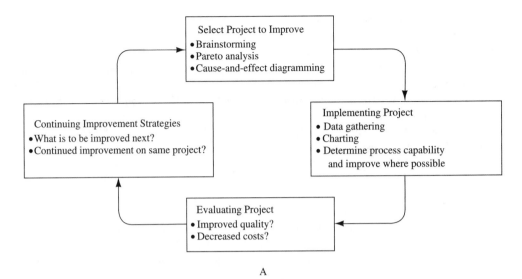

A

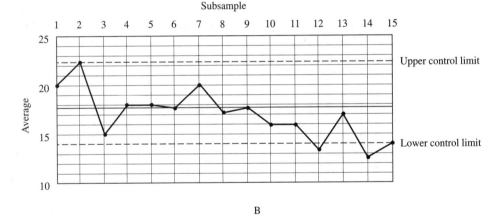

B

Figure 3.12 (A) The Shewhart cycle is the foundation on which SPC problem solving is structured. (B) An automobile's gas mileage is plotted over time.

3.4 Professionalism

Technicians must understand the ethics of the engineering profession they serve. The *International Society for Measurement and Control* (ISA) is one of the several professional societies in engineering with a written code of ethics. This excellently framed code, which specifically mentions the technician, follows (courtesy of International Society for Measurement and Control).

International Society for Measurement and Control Code of Ethics

PREAMBLE

As engineers, scientists, educators, technicians, sales representatives, and executives in an important and learned profession; and in order to safeguard public welfare; and to establish and maintain a high standard of integrity and practice; and as members of the International Society for Measurement and Control, we hold to these Articles:

ARTICLE I

Members shall hold paramount the safety, health and welfare of the public in the performance of their duties, and shall notify their employer or client and such other authority as may be appropriate where such obligations are abused.

Members shall hold in confidence facts, data and information obtained in a professional capacity, unless the release thereof is authorized by their employer or client, and shall not engage in fraudulent or dishonest business or professional practices.

ARTICLE II

Members shall perform services only in areas in which they are qualified by education or experience, and shall endeavor to maintain their professional skills at the state of the art. Members shall practice their profession in a manner which will uphold public appreciation of the services they render.

ARTICLE III

Members shall issue public statements only in an objective and truthful manner, and shall include all pertinent and relevant information in professional reports, statements and testimony. Members shall be honest and realistic in making estimates or in stating claims based on available data. Members shall offer honest criticism of work, and shall properly credit the contributions of others.

ARTICLE IV

Members shall act in professional matters for each employer or client as faithful agents or trustees, and shall not participate in any business association, interest or circumstances which influence, or appear to influence, their judgment or the quality of their services. Members shall accept compensation, financial or otherwise, from only one party for services on or pertaining to the same work, unless otherwise agreed to by all parties; and shall not give or accept, directly or indirectly, any gift, payment or service of more than nominal value to or from those having business relationships with their employers or clients.

ARTICLE V

Members shall use only proper solicitation of employments, and shall represent their abilities, qualifications, education, technical associations and professional registrations without exaggeration and in accordance with the laws of the locations in which they practice.

ARTICLE VI

Members shall pledge themselves to live and work according to the laws of man and to the highest standards of professional conduct, using their knowledge and skills to the benefit of all mankind.

(Adopted by the ISA Executive Board, October 17, 1986)

The Code of Ethics helps technicians and technologists to place their work in ethical perspective. *Ethics* is a vast and complicated subject, but may be best understood as a set of rules or laws such as the "Code of Ethics of the International Society for Measurement and Control." Ethical codes offer all who work in technical areas a standard to guide them in their day-to-day decisions—decisions that may affect hundreds, thousands, or even millions of people. Most of the professional engineering/technical societies listed in Appendix A have written codes.

The last fundamental principle is often overlooked by the paraprofessional and is important from a career standpoint. Many professional and technical societies not only recognize technicians and technologists as important members of the "technical team," but encourage their participation. The International Society for Measurement and Control and the Society of Manufacturing Engineers (SME) are outstanding examples of such interest in paraprofessionals. Both organizations support student chapters in the two-year colleges.

Professional organizations for engineering and technology must inform management of the need for and worth of those who contribute to manufacturing in this country. Sony chairman Akio Morita criticized American management in a 1989 *Newsweek* article. He stated:

> American industry and management pay too much attention to the bottom line, to the quarterly report, to fluctuations in the stock price. Businessmen who can make money quickly are very highly respected. But engineers or technicians who have to work in the factory are not so respected. Actually, these people are essential to industry. In Japan, engineers [and technologists] are highly respected: that's why they're willing to work on the factory floor.

See Appendix A for information concerning professional societies in engineering and technology. You may wish to participate in your college's student section or help to establish one if a student section does not exist.

Standardization

Standardization is one of the primary services offered by professional societies. Standards agreed upon by all, nationally and internationally, greatly increase efficiency and eliminate waste. When we measure by the same measurement systems (see Chapter 5), and agree on the dimensions of standard components, the whole society benefits.

The American National Standards Institute (ANSI) is the coordinating organization for America's voluntary national standards system. Founded in 1918, the ANSI federation consists of approximately 1000 companies, 30 government agencies, and 250 professional and technical organizations. ANSI services include

1. coordinating voluntary standards activities, so all parties involved may be optimally satisfied,
2. approving standards as American national (consensus) standards,
3. representing U.S. interests in international standardization, and
4. providing information and access to the world's standards.

There are currently 8,000 national standards, which include such varied items as dimensions, ratings, nomenclature, symbols, test methods, and performance and safety requirements. Decision makers are technical representatives from all user groups. These volunteers sit on committees and review new standards for approval. Users of these standards are manufacturers and distributors, retailers, banks, utilities, insurance companies, consumers, construction industries, and government agencies.

Certification

Certification is one way a technician may offer evidence of competence in a specific technology. Increasing specialization has encouraged the growth of certification, and many certification examinations are more specific than a particular technology (e.g., Certified Quality Technician [CQT]).

The National Institute for Certification in Engineering Technologies (NICET) offers certification to technicians and technologists who voluntarily apply for certification. Formal education and work experience are factors in certification, but the technician or technologist also must "sit for an examination" at an approved test site. Successful candidates are known as Certified Engineering Technicians (CETs). Many technical colleges offer certification tests to their graduates on campus and just prior to graduation.

Technologists or BSETs are qualified to sit for the Engineer in Training Examination (EIT) in 31 states. The EIT is a prerequisite to qualifying as a professional engineer (PE).

Increasing specialization has encouraged the growth of certification, and many certification examinations are even more specific than those that test for competence in an entire technology. For instance, there is a test for *Certified Quality Technician [CQT]* that fits within Industrial Engineering Technology.

Specific technical programs are often accredited by the Accreditation Board for Engineering and Technology (ABET). ABET evaluates the quality of engineering education by accrediting college curriculums in engineering and engineering technology.

Your technical or community college should have further information on specific programs accredited by ABET. Many technical colleges have not sought accreditation because of the expense of seeking it. These colleges also realize graduates of non-ABET accredited programs can still sit for technical certification exams and become certified upon passing.

The Society of Women Engineers (SWE) will be of interest to women technicians and technologists. SWE supplies information on the achievements of women engineers and the opportunities open to them. The organization also assists women to return to active work following temporary retirement. The society offers a periodical five times a year. Appendix A has contact information.

Problems

Section 3.1

1. Describe the A.S.E.T. and B.S.E.T. programs. What do the initials stand for? What is the central purpose of both programs?

2. How does a B.S. degree program differ from a B.S.E.T. degree program?

3. List (a) the key components of the five skills and knowledge that employers look for in technicians; and (b) the two important prerequisites to success in college.

4. What does the abbreviation *co-op* stand for? Who are the partners that form a co-op program and how does it work for the student who chooses to co-op?

5. Prepare a one-page résumé. Use the model offered in the subsection titled "Cooperative Education."

6. Check with the placement office for hardware and software that may be used to build an effective résumé.

7. Your technical education will be more meaningful if you can also work in a technically related field while attending school. Inquire at the engineering technology division office or at the college placement office about cooperative education programs and/or part-time employment opportunities in your field. List and describe the school/work employment offerings for your field at your college.

8. List the steps you would take to prepare for an interview. Also list two questions you might ask the interviewer.

9. Begin preparing a graduation portfolio. Use the DOT materials (problem 13) to begin your file. Place all course syllabi in your files as you progress through your engineering technology education, so you will not forget the content of your courses and the skills and knowledge you were exposed to.

10. Draw up a day-by-day schedule for one week, describing in-class times and regularly scheduled study periods. Submit this weekly schedule for the present college term to your instructor.

11. Prepare a plan of study to complete your graduation requirements, including the list of courses you will take each term to achieve your goal. Identify each course as math-science, communications or general, or technical (specific to your technical area). Check your plan with those of other students entered in the same technical program.

12. Memorize Table 3.1 and be prepared to be examined by the instructor at any time. *Hint:* The most frequently misspelled words in the table are 1, 3, 12, 16, 20, 21, 23, 24, 26, and 27.

Section 3.2

13. Using the DOT (Section 3.2) find three occupations of interest to you. Photocopy each position description (if photocopy machines are available) and submit it to your instructor. Or write a brief (one-page) synopsis of the three position descriptions, if your instructor prefers.

14. Using your college's library computer system (see Figure 3.7), do a search of one engineering subject of interest to you. Print out an abstract of one of the articles found. Submit the printout to your instructor.

15. Research the *Applied Science and Technology Index* in the library and find one journal article of interest to you. Locate the journal of interest, photocopy the first page of the article (if photocopy machines are available), and submit it to your instructor. Or write a brief (one-page) synopsis of the article, if your instructor prefers.

16. Ask the reference librarian at your college to help you find one reference work, for example, the DOT or an engineering handbook. Inquire as to how the librarian would access information in that particular reference source.

17. It is important that you can calculate your grade point average (GPA). The GPA is a good example of a weighted average. In this case the average depends on the number of credit hours for each course grade. Complete the table below by calculating the grade point for each course (example based on 4-point system: A = 4, B = 3, C = 2, D = 1, E or F = 0 points).

Course Credit Hours	Course Letter Grade	Calculated Grade Point
4	A	$4 \times 4 = 16$
2	D	_____
6	C	_____
3	E (or F)	_____

After calculating the grade points, add them and divide by the number of credits to find the GPA for the quarter. In equation form:

$$\text{GPA} = \frac{\Sigma \text{ grade points}}{n \text{ (credits)}} = \text{_____}$$

18. In problem 15, change the E to a B. What is the new GPA?

Section 3.3

19. Be prepared to list from memory the seven basic elements of problem solving.

20. Briefly describe a problem that you have encountered recently. List your alternatives and indicate the alternative selected as most appropriate. Provide a brief rationale for selecting that alternative.

Use the problem solution format demonstrated in Examples 3.1 and 3.2 to write a solution for one or more (instructor's determination) of Problems 21 through 25. These problems are typical of those found in industry. Note that a lot of information is absent, but that is what makes them problems! Students should feel free to add their own conditions or intermediate outcomes during the development of a solution.

21. After connecting an electronic circuit designed by an engineering supervisor, you find the output undesirable.

22. The production level on your assembly line is lower than usual. As shift supervisor, you hear a rumor that they are "out to get you."

23. You have been asked to design a simple hydraulic circuit. You have had only one course in hydraulics at the technical college and no experience on the job.

24. You are employed by an air-conditioning manufacturer. This morning the boss came to your office and asked that you travel to a large customer's location and troubleshoot a faulty system. Your plane leaves tonight and you are unfamiliar with that particular system.

25. You rarely see or speak with your manager. She avoids giving you feedback regarding your work—either praise or disapproval. Your annual evaluation interview is scheduled for next week.

Section 3.4

26. Attend at least one meeting of an engineering society at your college or in your community. Ask about cost of student membership, benefits of membership, and so on. Write a brief report describing your findings.

27. What does the acronym ANSI stand for? List two of the four services ANSI provides.

28. State one way—other than by possessing a college degree—that a technician may offer evidence for competence in his or her technology.

29. Read ahead in Chapter 4, attempting to use your calculator to work through the examples. Consult your operator's manual as you work the examples. Consult your instructor as to how you should show evidence of this assignment.

30. Through personal research write a one-page report on how a great person (as in the following list) overcame adversity or used good problem-solving methods: Marie Curie, Robert H. Goddard, George Washington Carver, Henry Ford, Thomas Edison, Shockley, Bardeen, and Brattain (inventors of the transistor).

Selected Readings

H. W. Wilson Company. *Applied Science and Technology Index.* Updated monthly.

U.S. Department of Labor. *Dictionary of Occupational Titles,* 4th ed. 1991. See occupations 003.161, 003.261, 005.281, 007.161, 008.261, 012.167, 012.267, and 019.281.

The technical journals in your college library, especially those recommended by your instructor and indexed in the *Applied Science and Technology Index.* Examples include *American Machinist, Aviation Week & Space Technology, Byte, Compressed Air, Electronics, Hydraulics and Pneumatics, InTech, Mechanical Engineering,* and *Quality Progress.*

Your college catalog or bulletin.

4

The Calculator

You must master the calculator to be successful in your studies—it is used throughout the engineering technology curriculum. Although the calculator reinforces the learning of mathematics concepts, it will not allow you to bypass a thorough understanding of those concepts. You will also enroll in mathematics courses, or the supporting mathematics may be included as an integral part of this "Introduction to Engineering Technology" course.

This chapter does address some of the necessary mathematical concepts you will need to fully understand the calculator. These "calculator essentials" include the *basics of algebra*—the number line and algebraic operations, *powers and roots,* and *powers of ten*—the language of large and small numbers. Appendix B contains the most important mathematical concepts in summary form. It is recommended that you memorize these concepts. The sign laws for algebraic addition, subtraction, multiplication, and division are extremely important. Many students have greater problems with the laws of addition and subtraction than with the laws of multiplication and division. Memorization of the rules of exponents will help you understand how powers of ten are used and interpreted with the calculator. Many of the examples that follow have more than one solution. Working through each solution will increase your understanding of the calculator and the underlying math principles. *It is most important to work through and understand all of the examples.*

4.1 Purchasing the Calculator

The calculator is an outstanding example of modern electronics and our ability to package large electronic circuits into small packages. Figure 4.1A is an enhanced (expanded) view of the integrated circuit (IC), or chip, that is the heart of the calculator.

When shopping for your calculator always ask for an owners' manual; it should be included in the purchase price of the calculator. The owner's manual

A

B

Figure 4.1 (A) An internal view of a calculator chip reveals the sophisticated architecture of the microcircuit. (B) The engineering technology student finds the scientific calculator indispensable during study sessions.

should contain at least 100 pages and include some business and engineering applications. *This chapter is designed to complement your owner's manual and not replace it.* Do not try to read the manual; instead, work examples in the manual and in this chapter while using the manual. The hands-on approach will increase your understanding of this vital instrument (Figure 4.1 B). During your career in technology you will have to read and understand operator's manuals for computers and

Table 4.1 Essential and Desirable Features of a Calculator*

Essential	Desirable
Sure key presses	Automatic power off
Minimum of 32 functions	Solar power
Scientific and engineering notation	Statistical functions
(Powers of Ten)	Hexadecimal conversion
Liquid crystal display	Unit conversions and angle/time
1000 hours battery life	conversions
3 levels of parentheses	Full equation onscreen
Deg/Rad/Grad keys	Programmable
Rectangular to polar conversion	Graphical

*Your technical instructor is the best resource in selecting the calculator most appropriate for your chosen technology.

other sophisticated equipment. These manuals will be much more complicated than your calculator's manual.

Consider the features listed in Table 4.1 when purchasing a calculator. The essential features are for calculators with single-line screens and without graphical capabilities. Many advanced calculators on the market have larger screens (up to 8 lines) with *equation operating systems* that allow users to enter numbers and operations in a simple, straightforward sequence. These calculators draw graphs on the larger display (Figure 4.2) and may often be programmed to automatically solve an equation. Selected *Examples* in this chapter will contain an additional *Solution* showing the keystroke sequence for the TI-85. These additions will help you to see the differences between the equation operating system on a large screen calculator, as contrasted with the keystroke sequence of a single-line calculator.

Hewlett-Packard developed one of the earliest handheld calculators, and its latest versions offer algebraic logic as well as the earlier reverse-polish notation (RPN) logic. The reverse-polish logic system has algebraic operations (functions) follow data entries: e.g., to add 3 to 4, input 3, input 4, and then press ⊞. Again, your owner's manual will help you blend the solutions in this chapter to the keystrokes necessary for your calculator. All of the examples in this text will use the algebraic logic system.

The keyboard is the logical place to begin a hands-on evaluation of the calculator. Most calculators today have sensitive and sure keyboards. Some keyboards have small keys; for these, a pencil eraser may be used to "press in" the key entry. Try using the pencil eraser when working many problems at one sitting. Use the *second-function* 2nd F key to press in functions above the keys on your calculator. Calculators with many functions, like the TI-85, have on-screen menu functions. Press the 2nd MODE keys to select the *MODE selection screen*. TI-85 users will need to use the first three lines of this screen to select Normal, Scientific, or Engineering Notation (Section 4.4); to select Floating or Fixed decimal point settings (Example 4.3); and to select the Degree mode for work in Chapter 6.

Also, evaluate the display and take time to learn what every item on the display is showing you. Scientific calculators have at least eight digits on the display. With scientific notation (Section 4.4) only 5 digits of accuracy is possible. *Your instructor will probably ask for answers to three digits of accuracy.* More information on rounding off is offered in Example 4.3 and in Section 7.3.

A   B

Figure 4.2   (A) The TI-85 and (B) TI-86 graphics calculators. (Courtesy Texas Instruments, Inc.)

4.2   Signed Numbers and Algebraic Logic

It is helpful to use the number line shown in Figure 4.3 to visualize the operations with signed numbers ($\pm$) on the calculator.

Figure 4.3   The concept of the number line must be understood to effectively use signed numbers.

EXAMPLE   Add the following signed numbers: -3, $+2$, $+6$, and -1.
4.1

Solution 1   The solution may be visualized as shown in Figure 4.4. The arrows in the sketch are vectors. A **vector** is defined as a quantity having both magnitude and direction and is represented by a directed line segment whose length represents magnitude and whose orientation represents direction.

Note: The direction of a vector on the number line during the addition process is to the right for plus ($+$) numbers and to the left for minus ($-$) numbers.

Solution 2   The paper-and-pencil solution shows parentheses around the vector values and plus signs (operation signs for algebraic addition) between the parentheses, to indicate algebraic addition rather than algebraic subtraction (the operation signs would be minus to indicate algebraic subtraction—see Appendix B and Problem 8).

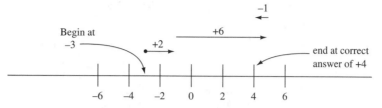

Figure 4.4 Vectors aid in visualizing the addition of signed numbers.

$$(-3) + (+2) + (+6) + (-1) = +4$$

or change the sign of the numbers first, as follows:

$$-3 + 2 + 6 - 1 = +4$$

and we are back to the original problem. This solution shows how operation signs differ from the signs of a vector value.

Note: Why use two signs when one will do? Always convert multiple signs by

$+ + = +$	i.e., $+(+2) = +2$
$- - = +$	i.e., $-(-2) = +2$
$- + = -$	i.e., $-(+2) = -2$
$+ - = -$	i.e., $+(-2) = -2$

Solution 3 Using the calculator, the keystrokes are as follows (boxed items represent the calculator keys):

Keystroke	Comments
3 $\boxed{+/-}$	Enter a -3 to begin by using the change-sign key $\boxed{+/-}$
$\boxed{+}$	Operation sign to add
2	The $+$ sign for the 2 is automatically entered by the calculator
$\boxed{+}$ 6	Add $\boxed{+}$ 6
$\boxed{-}$	Operation sign to subtract
1	1 is to be subtracted from the previous operations
$\boxed{=}$	Will complete the calculation with the 4 ($+4$) displayed as the correct answer

Solution 4 Using the TI-85, the keystrokes are as follows: $\boxed{(-)}$ 3 $\boxed{+}$ 2 $\boxed{+}$ 6 $\boxed{-}$ 1 $\boxed{\text{ENTER}}$

This brief example of adding signed numbers and the change-sign $\boxed{+/-}$ operation will help you to transfer your more flexible basic mathematical skills (paper-and-pencil skills) to the more rigid format and operating system of the calculator.*

---

*Note the difference between the change sign key ($\boxed{+/-}$ or $\boxed{(-)}$) and the operation key $\boxed{-}$ on the right-hand side of the calculator keyboard.

Table 4.2   The Algebraic Order of Operations*

Level	Operations
1	Single-variable functions calculated on entry such as $\boxed{\text{SIN}}$, $\boxed{1/x}$, $\boxed{x^2}$, and $\boxed{\%}$
2	The power keys such as $\boxed{Y^x}$ and $\boxed{\sqrt[x]{Y}}$
3	Multiplication $\boxed{x}$ and division $\boxed{\div}$
4	Addition $\boxed{+}$ and subtraction $\boxed{-}$
5	The equals key $\boxed{=}$ completes all pending operations

*Consult your owners manual for further definition. The TI-85, for instance, offers a more complete order of operations known as the *Equation Operating System* (EOS™).

Remember, to be a good problem solver, you must be able to predict the appropriate range of the answer the calculator gives you. Do not depend on the calculator too much. Have confidence in your basic math and engineering skills to identify input errors to the calculator.

Move next to the operating system for the calculators covered in this chapter—the *algebraic logic system* (ALS), often referred to simply as the algebraic method. A calculator based on the algebraic method performs all operations in an established order. Refer to Table 4.2 to understand the order for all operations.

Parentheses may be inserted to modify the order of operations. It may help you to memorize the use of parenthesis and the algebraic order of operations if you remember the sentence "**P**lease **E**xcuse **M**y **D**ear **A**unt **S**ally." The first letter in each word stands for **P**arentheses, **E**xponents (and other single-variable functions), **M**ultiplication, **D**ivision, **A**ddition, and **S**ubtraction.

EXAMPLE 4.2

Solve the following expression:

$$2 \times 4 - 6 \div 2$$

Note that the signs of the numbers are all positive. The minus sign is an operation sign.

Solution 1   Your paper-and-pencil solution must use the algebraic hierarchy. Parentheses are used so that any multiplication and division are accomplished first:

$$(2 \times 4) - (6 \div 2) = 8 - 3 = 5$$

Solution 2   To test your single-line screen calculator's ALS, key in the first expression without the parentheses. As you enter the data, watch the display to see when the calculator accomplishes each operation.

Keystroke	Display	Comments
2	2	The +2 is entered.
$\boxed{x}$	2	Operation sign to multiply (You may change operations as many times as you wish and your calculator will use only the last one entered.)

4	4	The +4 is entered.
⊟	8	A subtraction operation is entered and the 4×2 operation is completed.
6	6	Note the 6 is not subtracted in order to allow multiplications and divisions to be carried out first.
⊡	6	Operation sign to divide.
2	2	The $6 \div 2$ operation is not yet completed, pending further operations.
⊟	5	Because ⊟ was pressed, there was no quotient of 3 displayed; it was completed automatically in the ALS, then the 3 was subtracted from the 8.

Parentheses keys are present on scientific calculators to allow you to establish your own requirements for solving an algebraic expression—in other words, to allow you to bypass the programmed algebraic hierarchy.

EXAMPLE 4.3

A vernier dial caliper, shown in Figure 4.5, is used to measure a one-half inch part. The photo is to scale and you can confirm the part is one-half inch by measurement (try this with a ruler). You may also confirm the caliper reading by using the bottom-vernier scale (the top-vernier scale is in millimeters and the movable jaw is positioned a little above 12.5 mm). The movable jaw is also positioned near the five-tenths inch mark on the lower scale. Simply add the reading on the dial indicator, 0.003, to the 0.5 and 0.503 in. is the result.

Precision measurements often vary. Some reasons for different readings are using more or less force on the jaws (operator touch), reading of the dial from different angles (parallax), and slop in the dial caliper itself (instrument precision). Section 7.2 discusses measurement errors.

In this case, the operator takes five measurements at the same location on the part. The measurements in inches are

$$X_1 = 0.503 \text{ in.}, X_2 = 0.499 \text{ in.}, X_3 = 0.500 \text{ in.}, X_4 = 0.500 \text{ in.}, X_5 = 0.502 \text{ in.}$$

Find the average or mean ($\bar{X}$), of the measurements.

Solution 1

$$\bar{X} = \frac{\Sigma X}{n} = \frac{X_1 + X_2 + X_3 + X_4 + X_5}{\text{number of measurements}}$$

$$= \frac{0.503 + 0.499 + 0.500 + 0.500 + 0.502}{5}$$

$$\bar{X} = \frac{2.504}{5} = 0.5008$$

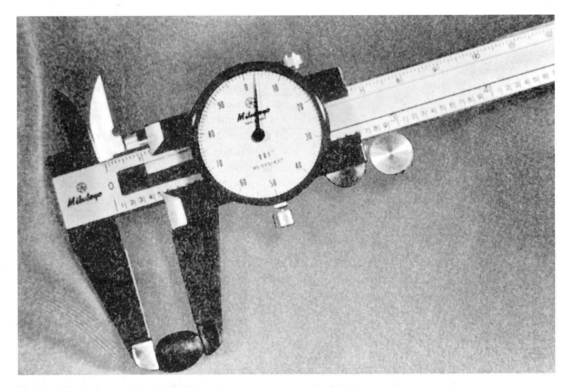

Figure 4.5   The vernier dial caliper is used to make quick and fairly accurate measurements in industry. (Courtesy National Institute of Standards and Technology)

The final answer is rounded off to the accuracy of the vernier caliper, or 0.501 in., close to the manufacturing target of 0.500 in. Accuracy is discussed in Section 7.3. The accuracy of the gauge in this example is three significant figures, meaning that the operator should not try to estimate the measurement beyond the least significant figure on the instrument (the dial on the vernier caliper). Since many gauges used in industry are accurate to three significant figures, *this text will require that all answers be accurate to three significant figures.*

Note: Do not confuse the number of decimal places with the number of significant figures in an answer. The number of significant figures is established by the ability to read a measurement to a certain number of digits of accuracy, regardless of the position of the decimal point. Refer to your operator's manual to select floating point or to *fix* the decimal point to a selected number of digits right of the decimal. On the TI-85 press [2nd] [MODE] and use the arrow keys to move to the second line of the MODE screen.

Solution 2   The calculator solution involves the following keystrokes:

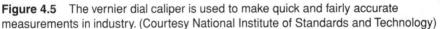

[(] 0.503 [+] 0.499 [+] 0.500 [+] 0.500 [+] 0.502 [)] [÷] 5 [=] 0.501 in.

To attempt Solution 2 without using parentheses would have resulted in an answer quite different from the correct average.

$$0.503 \boxed{+} 0.499 \boxed{+} 0.500 \boxed{+} 0.500 \boxed{+} 0.502 \boxed{\div} 5 \boxed{=} 2.10 \text{ in.}$$

Can you say why this calculation is incorrect? Refer to Table 4.2, The Algebraic Order of Operations.

Solution 3　Parentheses also must be used with the TI-85. Consult *The Equation Operating System* in the TI-85 *Guidebook.* Try the calculation with and without parentheses in order to assure yourself of the need for parentheses.

Nested parentheses can also be calculated, as the following examples demonstrate.

Note:　You must check your operator's manual for the number of levels the calculator can handle during a single calculation.

EXAMPLE
4.4

Calculate the following expression containing two levels of parentheses.

$$3[-2(4 + 7 - 3)] =$$

Solution 1　The paper-and-pencil solution involves a process of working from the innermost level of parentheses out:

$$(4 + 7 - 3) = 8$$

The multiplication sign is not shown, but

$$[-2 \times 8] = -16$$

and, again, the implied multiplication completes the problem:

$$3 \times [-16] = -48$$

Solution 2　The calculator solution is as follows:

$$3 \boxed{\times} \boxed{(} 2 \boxed{+/-} \boxed{\times} \boxed{(} 4 \boxed{+} 7 \boxed{-} 3 \boxed{)} \boxed{)} \boxed{=} -48$$

Note:　Most calculators will not automatically insert a multiplication operation $\boxed{\times}$ between parentheses or between a factor and the parentheses. The TI-85 does recognize implied multiplication and the keystrokes for this problem are

$$3 \boxed{(} \boxed{(-)} 2 \boxed{(} 4 \boxed{+} 7 \boxed{-} 3 \boxed{)} \boxed{)} \boxed{\text{ENTER}}$$

EXAMPLE
4.5

The resistance (R_2) of a wire varies according to the temperature (t) as

$$R_2 = R_1[1 + \alpha(t_2 - t_1)]$$

The factor (α) is found in tables from engineering handbooks. Alpha (α) varies by material type. Find the resistance of an aluminum wire ($\alpha = 0.004$) at 100°C (t_2) if the resistance (R_1) is 20 Ω at 10°C (t_1).

Solution The resultant algebraic expression after all values are plugged into the equation is

$$R_2 = 20[1 + 0.004(100 - 10)]$$

and the calculator keystrokes are

20 [×] [(] 1 [+] 0.004 [×] [(] 100 [−] 10 [)] [)] [=] 27.2 (Ω)

There is also a difference, mathematically, between dividing a number into a sum (**polynomial**) or a product (**monomial**), as the following example will demonstrate.

EXAMPLE 4.6 Divide 6 into the sum of $2 + 5 + 4$ and the product of $2 \times 5 \times 4$.

Solution 1

$$\frac{2 + 5 + 4}{6} = \frac{11}{6} = 1.83$$

Solution 2

[(] 2 [+] 5 [+] 4 [)] [÷] 6 [=] 1.83

Parentheses must be used with the polynomial. In this example the algebraic hierarchy would allow only the 4 (the last number in the polynomial) to be divided by the 6 if the parentheses had not instructed it differently. Confirm this with your calculator by using the following keystrokes:

2 [+] 5 [+] 4 [÷] 6 [=] 7.67 (this is not the correct answer, as above)

Now for the monomial solution:

Solution 1

$$\frac{2 \times 5 \times 4}{6} = 6.67$$

Note that the 6 could be divided into any number (term) in the numerator or could be divided into the final product. Therefore, no parentheses need be employed with the calculator.

Solution 2

2 [×] 5 [×] 4 [÷] 6 [=] 6.67, or 2 [÷] 6 [×] 5 [×] 4 [=] 6.67

In the case of the several products being in the denominator, such as $\frac{6}{2 \times 5 \times 4}$, you cannot use 6 ÷ 2 × 5 × 4. But you may use 6 ÷ 2 ÷ 5 ÷ 4. The 6 is to be divided by all three monomial factors. Using parentheses makes the calculation simpler:

6 ÷ (2 × 5 × 4) = 0.15

When in doubt, use the parentheses to calculate fractions involving signed numbers.

It is useful to recognize the different techniques used in working with monomials and polynomials (refer to Appendix B, "Illegal Operations").

4.3 Powers and Roots

When a number is squared, it is multiplied by itself. It is interesting to note that the result may be visualized as a square or as a two-dimensional figure.

EXAMPLE 4.7

An auditorium floor is square and is 40 ft on one side. It is to be covered with carpet. Find the area in square feet.

Solution 1   First, draw a picture of the problem, so you can visualize it (Figure 4.6). Use the x^2 key to find the area in square feet:

40 x^2 = 1600 ft^2

Solution 2   You may also use the more powerful "power key" y^x to confirm the previous result:

40   y^x 2   = 1600 (1600 ft^2)
↑   ↑
the   the
base   power

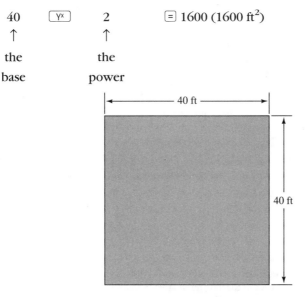

Figure 4.6   Drawing of the square auditorium floor to be carpeted. Note the standard dimensioning.

The number 2 in X^2 is called an exponent. The **exponent, or power,** indicates how many times a base (e.g., the X on the $\boxed{x^2}$ key) is to be multiplied. If the exponent is three, the result may be visualized as a cube (X^3).

Solution 3   Using the TI-85, the keystrokes are as follows:

40 $\boxed{\wedge}$ 2 $\boxed{\text{ENTER}}$

EXAMPLE
4.8

The auditorium in Example 4.7 is a perfect cube, meaning that the height of the ceiling is 40 ft (Figure 4.7). Solve for the total volume contained within its walls.

Solution   Again, the picture is drawn to visualize the problem (Figure 4.7). Use the power key $\boxed{y^x}$ to cube the 40 (some calculators have an $\boxed{x^3}$ key):

40 $\boxed{y^x}$ 3 $\boxed{=}$ 64 000* (64 000 ft^3 [ft $\times$ ft $\times$ ft = ft^3])

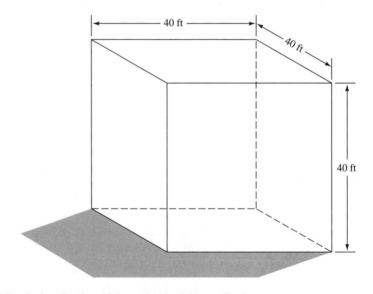

Figure 4.7   A visualization of the volume of the auditorium.

---

*Outside the United States, the comma is used as a decimal marker (i.e., 23,456 = 23.456). To avoid confusion, recommended international practice calls for separating the digits into groups of three, counting from the decimal point toward the left and the right and using a small space to separate the groups. The spacing is not necessary for only four digits. In the example above 23,456 = 23 456. This text follows the recommended practice.

A root is calculated by using the inverse of the universal power key ⬜INV ⬜Yˣ or by directly using the universal root key ⬜ˣ√Y. To demonstrate the root function, find the cube root of the answer obtained in Example 4.8:

$$\sqrt[3]{64\ 000} \text{ is} \quad 64\ 000 \quad \boxed{ˣ√Y} \quad 3 \quad \boxed{=} \quad 40 \ (40 \text{ ft})$$

$$\uparrow \qquad\qquad\qquad \uparrow$$

$$\text{the} \qquad\qquad\qquad \text{the}$$

$$\text{base} \qquad\qquad\qquad \text{root}$$

It is useful for the technician to understand that a **root** is a **fractional power.** In other words, $\sqrt[x]{Y} = Y^{1/x}$ (in the previous example, $[64\ 000 \text{ ft}^3]^{1/3}$). Use the **reciprocal key** $\boxed{1/x}$ to demonstrate this calculation on the calculator.

$$64\ 000 \text{ ft}^3 \quad \boxed{Yˣ} \quad 3 \quad \boxed{1/x} \quad \boxed{=} \quad 40 \ (40 \text{ ft})$$

$$\uparrow \qquad\qquad\qquad \uparrow$$

$$\text{the base} \qquad\qquad \text{inverts}$$

$$\text{the power}$$

For the TI-85, use: $64\ 000 \ \boxed{\wedge} \ 3 \ \boxed{\text{2nd}} \ \boxed{x^{-1}} \ \boxed{\text{ENTER}}$.

Refer to Appendix B, "Algebraic Rules." Under the Rules of Exponents find $a^{-x} = \frac{1}{a^x}$. Translated, this means that $X^{-1} = \frac{1}{x^1} = \frac{1}{x}$.

Prior to the development of the modern engineering calculator, the calculation of powers and roots was laborious and often required the use of logarithms. You will be required to solve expressions involving exponents. Consult your operator's manual to become more experienced with this important feature of the engineering calculator. Also, refer to a summary of the rules of exponents in Appendix B.

4.4 Using Scientific Notation

Technical personnel must work with very large and very small numbers—numbers that many other people would find impossible to grasp or work with. Often these numbers will not even fit on the display of the calculator because of the large number of zeros necessary to hold the decimal place. Some shorthand method is needed. This shorthand method is scientific notation. **Scientific notation** uses powers of ten, and numbers in scientific notation are always shown as *a number between one and ten times a power of ten.* Powers of ten keep track of a decimal point, just as zeros can. See Table 4.3 to understand this scientific shorthand.

You can practice entering the numbers in Table 4.3 into your calculator. After entering the number in the left-hand column, which is in standard form (floating point mode for the calculator), press $\boxed{=}$. Then use an appropriate operation key (consult your operator's manual) to convert the number to scientific notation. If your calculator possesses an engineering mode, acquaint yourself with this function also.

Table 4.3   Relationships Between Some Powers of Ten and the More Familiar Decimal Numbers

1 000 000	1×10^6	$(10^6)*$	Read as *one million*
100 000	1×10^5	(10^5)	Read as *one hundred thousand*
10 000	1×10^4	(10^4)	Read as *ten thousand*
1000	1×10^3	(10^3)	Read as *one thousand*
100	1×10^2	(10^2)	Read as *one hundred*
10	1×10^1	(10^1)	Read as *ten*
1	1×10^0		Read as *one* (any number to the zero power is one)
0.1	1×10^{-1}	(10^{-1})	Read as *one tenth* (a negative exponent becomes positive when inverted therefore $10^{-1} = 1/10$, or 0.1)
0.01	1×10^{-2}	(10^{-2})	Read as *one hundredth*
0.001	1×10^{-3}	(10^{-3})	Read as *one thousandth*
0.000 1	1×10^{-4}	(10^{-4})	Read as *one ten-thousandth*
0.000 01	1×10^{-5}	(10^{-5})	Read as *one hundred-thousandth*
0.000 001	1×10^{-6}	(10^{-6})	Read as *one millionth*

*The power of ten reflects the number of places the decimal point is moved to reach the zero power or the unit location. The expressions in parentheses represent the alternative method of writing powers of ten when the only significant figure (number) is one.

Engineering notation converts numbers in scientific notation to numbers larger than one and having a power of ten evenly divisible by three. Engineering prefixes are always expressed in multiples of three.

To enter a number already in scientific notation into the calculator, simply press the ⎡EE⎤ or ⎡EXP⎤ key. Try this by entering the numbers in the middle column of Table 4.3 (e.g., 1×10^6) into the calculator, press equals, and then convert to floating point and engineering modes. Please realize that when you press the ⎡EXP⎤ key that the base 10 of the power of ten is entered. Many students press the ⎡EXP⎤ key and then press ⎡x⎤ 10. If you make this common mistake, the results on the screen will be ten times too large.

The TI-85 has a **MODE** menu, used to convert to normal, scientific, or engineering notation. Press ⎡2nd⎤ ⎡MODE⎤ to display the mode menu. The default mode for the first row is highlighted. The selected mode is **NORMAL.** Press the right arrow key ⎡>⎤ to highlight the scientific notation (SCI) or engineering notation (ENG). Press the ⎡ENTER⎤ key when the desired notation is highlighted. Press the ⎡EXIT⎤ or ⎡CLEAR⎤ keys to return to the **HOME** screen.

Note: Some scientific calculators will not enter a power of ten when the numerical quantity is one (i.e., 1×10^3) unless you first enter a 1. Others will allow you to press ⎡EE⎤ or ⎡EXP⎤ first and simply enter the exponent. This is why it is imperative to practice with the calculator you have purchased.

EXAMPLE
4.9

Enter the numbers 1 000 000, 9820, and 0.000 605 in scientific notation into the calculator. Use the ⌈EE⌉ key to convert the numbers to scientific notation.

Solution   As the decimal point is moved to the left, making the number larger, the power of ten becomes more positive. Moving the decimal to the right results in a more negative exponent. The caret (^) indicates the position of the decimal point after conversion.

To Enter the Number	Press Calculator Keys	The Display Reads	
1ˆ000 000	1 EXP 6	1.0	06
9ˆ820	9.82 EXP 3	9.82	03
0.0006ˆ05	6.05 EXP +/− 4	6.05	−04

Multiplication and division with powers of ten is easily accomplished, as the following examples demonstrate.

EXAMPLE
4.10

Evaluate the expression: $8.7 \times 10^8 \times 9.3 \times 10^6$

Solution 1   The paper-and-pencil solution with powers of ten involves two distinct steps: solving the **numbers portion** (use the calculator for this step) and solving the **power of ten portion** of the problem. First the number portion is calculated:

$8.7 \times 9.3 = 80.9$

The powers of ten are calculated as follows:

$10^8 \times 10^6 = 10^{(8+6)} = 10^{14}$

Combining the answers from the preceding two steps results in the final answer:

80.9×10^{14}

Note: $10^x \times 10^y = 10^{(x+y)}$ To multiply powers of ten, add the powers. Refer to Appendix B, "Rules of Exponents": $a^x \times a^y = a^{x+y}$.

Solution 2   The calculator solution is

8.7 ⌈EXP⌉ 8 ⌈x⌉ 9.3 ⌈EXP⌉ 6 ⌈=⌉ 8.091×10^{15} (or 80.9×10^{14})

Solution 3   Using the TI-85, the keystrokes are as follows:

8.7 ⌈EE⌉ 8 ⌈x⌉ 9.3 ⌈EE⌉ 6 ⌈ENTER⌉ 8.09 E 15

It appears as if the calculator solution is different from that of the paper-and-pencil solution (80.9×10^{14}). It is not, due to the different placement of the decimal point.

You will find the movement of the decimal point and the subsequent change in the power of ten easier if you remember the following rules:

If you make the number larger by moving the decimal point to the right, then you must balance that action—make the power of ten smaller or more negative, e.g.,

$$8.743\ 69 \times 10^{11} = 874\ 369 \times 10^{(11-5)} = 874\ 369 \times 10^6$$

The number is made larger and the power of ten made smaller.

If you make the number smaller by moving the decimal point to the left, then you must balance that action—make the power of ten larger or more positive, e.g.,

$$35\ 144.4 \times 10^2 = 3.514\ 44 \times 10^{(2+4)} = 3.514\ 44 \times 10^6$$

The number is made smaller and the power of ten made more positive.

The above rules can be applied to Example 4.10:

$$8.10 \times 10^{15} = 81 \times 10^{(15-1)} = 81 \times 10^{14}$$

The number is made larger and the power of ten must be made smaller or more negative.

EXAMPLE 4.11

Evaluate the expression $3.66 \times 10^{-3} \div 2.85 \times 10^8$

Solution 1   Again, there are two distinct steps involved in the paper-and-pencil solution with powers of ten. First solve the numbers portion:

$3.66 \div 2.85 = 1.28$ (accurate to three significant figures)

Then solve the powers of ten portion of the problem:

$$10^{-3} \div 10^8 = 10^{(-3-8)} = 10^{-11}$$

Combining the answers from the previous two steps results in the final answer:

$$1.28 \times 10^{-11}$$

Note: $\dfrac{10^x}{10^y} = 10^{(x-y)}$

To divide powers of ten, subtract the powers. Refer to Appendix B, "Rules of Exponents": $a^x \div a^y = a^{x-y}$.

When moving the power of ten from the denominator to the numerator, simply change the sign of the exponent. You may check this by using the $\boxed{1/x}$ key on the calculator—key in 1×10^{12} and press $\boxed{1/x}$. The result is the reciprocal, or 1×10^{-12}.

Solution 2   The calculator solution is

$$3.66 \ \boxed{EXP} \ \boxed{+/-} \ 3 \ \boxed{\div} \ 2.85 \ \boxed{EXP} \ 8 \ \boxed{=} \ 1.28 - 11 \ (\text{or } 1.28 \times 10^{-11})$$

Solution 3   For the TI-85, use $3.66 \ \boxed{EE} \ \boxed{(-)} \ 3 \ \boxed{\div} \ 2.85 \ \boxed{EE} \ 8 \ \boxed{ENTER}$

Note: It is inappropriate to report a number expressed as a power of ten as it appears on the calculator display. Always write the expression as a number times a power of ten (not $1.28 - 11$, but 1.28×10^{-11}) in your report. Some instructors will allow the TI-85 display. In this case, $1.28 \text{ E} - 11$.

EXAMPLE
4.12

Confirm the results of the following calculations using your calculator.

$$1.5 \times 10^{-8} \times 2.38 \times 10^{-3} = 3.57 \times 10^{-11}$$

$$\frac{620 \times 10^5 \times 3 \times 10^{-5}}{848 \times 10^6 \times 5.08 \times 10^3} = 4.32 \times 10^{-10}$$

Solution   Use your calculator to confirm the above.

Next, try to divide 10^8 by 4 on your calculator.

Solution

$$\frac{10^8}{4} = 2.5 \times 10^7$$

To see this rather surprising answer, rewrite the problem as

$$\frac{1 \times 10^8}{4}$$

Remember to always calculate the number part of a power of ten expression and the power of ten part separately. The calculator will solve the problem correctly if the expression is entered correctly.

When powers of ten are added and subtracted a different rule is required, for example:

$$8.3 \times 10^8 - 6.74 \times 10^9 = -5.91 \times 10^9$$

Note: When powers of ten are added and subtracted with paper and pencil, the powers of ten must be equal and the common power of ten is carried over (e.g., $8.3 \times 10^8 - 67.4 \times 10^8 = -59.1 \times 10^8$). The calculator will complete the operation correctly if you, the operator, enter it as written and in its original form.

EXAMPLE
4.13

The total area of the foundation for a large building is 90 000 square feet. The building can exert no more than 2400 pounds per square foot on the subsoil strata (Figure 4.8). What is the maximum design weight for the building in pounds?

Solution   To calculate the maximum weight simply multiply the area in square feet (ft^2) by the maximum allowable load in pounds per square foot (lb/ft^2).

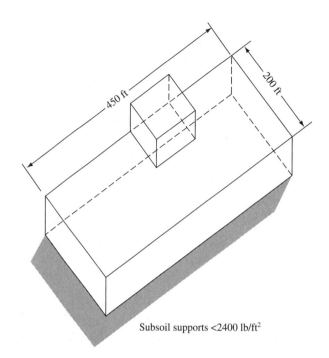

Figure 4.8   Isometric drawing of large building.

$$90\ 000\ \text{ft}^2 \times 2400\ \frac{\text{lb}}{\text{ft}^2} = 216\ 000\ 000\ \text{lb}$$

(this may be displayed on the 10-digit calculator)

$$= 2.16 \times 10^8\ \text{lb}$$

(in scientific notation mode)

Note that the units (ft^2) cancel in the solution's equation. The converting of units, a most powerful tool to the technician, is discussed at length in Chapter 5.

If you had entered the numbers into the calculator in scientific notation—a must for larger numbers—the keystrokes would be as follows:

9 (EXP) 4 (x) 2.4 (EXP) 3 (=) 2.160 08 (or 2.16×10^8 lb)

Converting to engineering notation, the display reads: 216×10^6 lb.

To convert to engineering notation with the TI-85, enter:

216 000 000 (2nd) (MODE), right arrow key to (ENG), press (ENTER), press (EXIT). Complete the calculation on the **HOME** screen by pressing (ENTER). The answer appears as 216 E6.

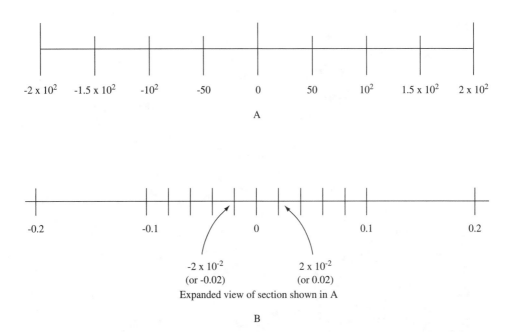

Figure 4.9 The number line in (A) is expanded near zero (B) to enable you to locate the very small numbers −0.02 and +0.02.

To further familiarize you with the relative values of the powers of ten, the following example with the number line is offered. The example clearly demonstrates how important it is not to confuse the sign of the number with the sign of the exponent.

EXAMPLE 4.14

Locate the following powers of ten on the number line.

$$2 \times 10^2, -2 \times 10^2, 2 \times 10^{-2}, -2 \times 10^{-2}$$

Solution Refer to Figure 4.9.

4.5 Graphing with the Calculator—"A Picture Is Worth a Thousand Words"

You will see many graphs during your technical education. Graphs allow you to visualize how two quantities change with each other, so you can predict how they will change in the future. In Figure 4.10 the N.C. technician depends on a graph depicting relationships between tool life and feed rate (right side of photo).

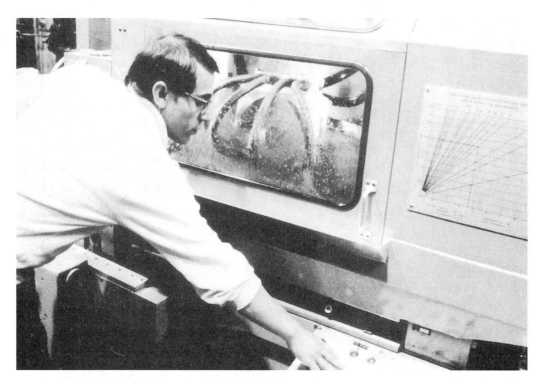

A

Figure 4.10   The operator of a high-speed numerically controlled lathe (A) predicts tool life
by viewing a graph similar to tool life vs. cutting speed (B). (Courtesy G.E. Superabrasives)

EXAMPLE   The current flow (I) through a load resistance (R) is directly proportional to the volt-
4.15   age drop across the resistance (Figure 4.11). This relationship is described by Ohm's
law, where

$$I = \frac{E}{R}$$

Graph the change in current due to a change in voltage of 0 to 10 V. Use a constant resistance of 25 000 ohms (25 kΩ).

Solution   Refer to Figure 4.12. The voltage (the *independent variable*) is being varied to cause the current change; hence, it is placed on the X, or horizontal, axis. The
calculator may be used to establish each of the plotted points (circled) by using the
memory key $\boxed{\text{STO}}$ or $\boxed{\text{X→M}}$. Because memory keys vary from calculator to calculator,
you should refer to your operator's manual to determine which keys store and re-
call data.

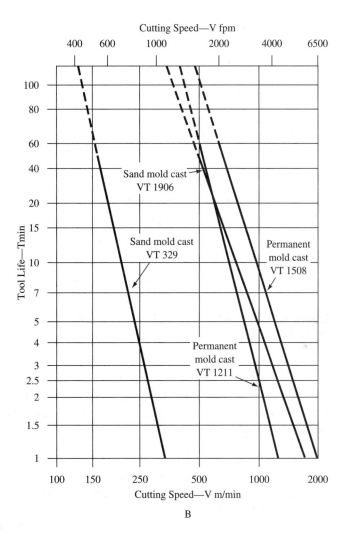

Figure 4.10 (continued)

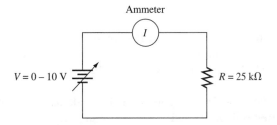

Figure 4.11 Schematic of circuit used to graph Ohm's law relationship. The voltage source is variable over a 10 V range. The current varies proportionally.

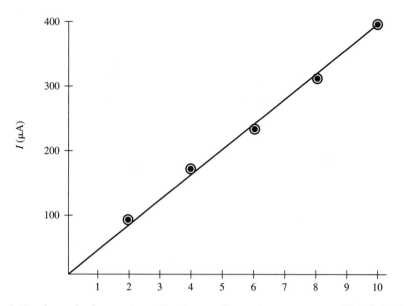

Figure 4.12 A graph of current resulting from voltage change across a 25 000 (25 kΩ) resistance.

For the example, we will use the terms *store* STO and *recall* RCL. Store the constant resistance ($R = 25$ kΩ) and repeatedly divide the constant value into the voltage steps of 0, 2, 4, 6, 8, and 10 V. The *dependent variable's* points are plotted on the previously graduated *Y* axis. The engineering notation mode is used for the example.

The calculator keystrokes for the first two calculations are
25 EXP 3 STO (stores 25 000 in memory, clears the display)
0 ÷ RCL = (answer is 0)
2 ÷ RCL = (answer is 80 −06, or 80×10^{-6}, or 80 microamperes [μA])

Continue to calculate and confirm the remaining plotted points. Problem 77 shows a solution for this example using the **TI-85.**

EXAMPLE
4.16

The current flow through a load (resistance) is inversely proportional to the resistance of the load. Graph the change in current due to a resistance change of 0 to 50 000 Ω (50 kΩ). Use a constant voltage of 120 V.

Solution   Refer to Figure 4.13. The calculator may be used to establish each of the plotted points (circled) by using Ohm's law.

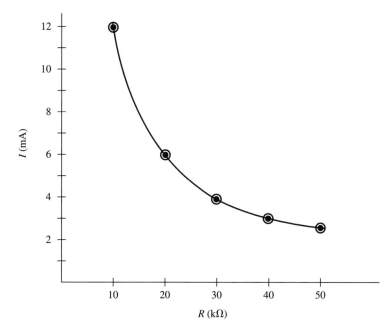

Figure 4.13 A graph of current change resulting from a resistance change. The constant voltage is 120 V.

$$\text{Current } (I) \;=\; \frac{\text{voltage } (V)}{\text{resistance } (R)} \;=\; \frac{120\text{ V}}{0-50\text{ k}\Omega}$$

$$= 120/0$$

The error flag on the calculator is displayed. (Dividing by 0 is undefined, but technicians think of this value as infinitely great. Plot the first point asymptotic [close to, but never touching] to the top of the Y-, or vertical, axis.)

$$I = 120/10\ 000 = 0.012 \text{ or } 12 \times 10^{-3} \text{ A, or } 12 \text{ mA}$$

$$I = 120/20\ 000 = 6 \text{ mA}$$

At this point, use the memory function on the calculator. Store the constant (120 V) and continue to calculate and confirm the points plotted on the graph in Figure 4.13.

The function graphed in Example 4.16, an inverse function, shows that the current changes much faster at the lower resistance values. If you examine the ohmmeter scale on a multimeter, you will find that the scale is not linear, but nonlinear. Further examination will reveal that the scale changes much the same as the graph of Figure 4.13.

EXAMPLE
4.17

The computer's operating system is based ultimately on a binary system (see Chapter 9). In a **binary system** there are only two possible states (i.e., a 0 and a 1) for each

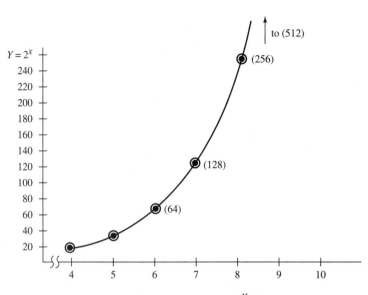

Figure 4.14 A graph of the exponential function $Y = 2^X$. The figures in parentheses are the actual Y values of the points plotted on the graph and represent typical memory sizes quoted by computer manufacturers.

"bit" of information. Graph the function 2^X to ascertain how memory amounts are determined for computer systems.

Solution   Refer to Figure 4.14 and use your calculator to confirm the points plotted on the graph. Observe that the ordinate (Y) value of the points plotted on the graph are typical memory figures quoted by computer manufacturers. Note: 1K of random accessory memory (RAM) consists of 8192 bits (1K of RAM = 1024 bytes, or 2^{10} bytes, and 8 bits make up a byte). Continue your evaluation of 2^X to find the value of X equaling 8192 bits.

Any good salesperson will tell you, "If you can paint a picture in your customer's mind, you stand a better chance of selling your idea or product." The three preceding examples demonstrate how the calculator can be used to assist in graphing quantities to visualize their changing relationship. More information on graphing is included in Chapter 7, "The Laboratory."

Graphing Calculators

Several "friendly" and easy-to-use inexpensive graphing calculators are available. Even though they cost more than less sophisticated calculators, graphing calculators have more functions, making them useful further into your career. Purchasing one of these powerful calculators might be a wise investment for the future. One caution, however, is that a graphing calculator takes more time to learn and more practice with the owner's manual.

If you have a graphing calculator you will want to try to graph Example 4.17. The following keystrokes taken from Texas Instruments' TI-85, will illustrate:

> Press GRAPH. Press f1 to display the <y(x)=> (function) editor. Enter 2 ∧ and then press f1 again to enter the x (function variable). Press 2nd f2 (<RANGE>), then enter the following: <XMin = 4> <xMax = 10> <xScl = 1> <yMin = 0> <yMax = 260> (or 520, depending on precision desired) and <yScl = 20>. Then press f5 (<GRAPH>) to display the graph. Press f4 (<TRACE>) to view the (blinking) cursor location and the corresponding x and y values at the bottom of the screen. The cursor may be moved with the arrow keys.

Consult your owner's manual for the exact procedure. Some of the functions are on screen, denoted by <> symbols.

EXAMPLE 4.18   Fold a sheet of 8 1/2 × 11 inch paper to form a box (Figure 4.15). Calculate the volume of your result by solving the equation

$$V = (L - 2X)(W - 2X)X$$

Using the TI-85, calculate the maximum volume possible for such a box.

Solution   Press GRAPH. Press f1 to display the <y(x)=> editor. Enter the expression to define the function y1. Use the equation for Y, in the previous paragraph,

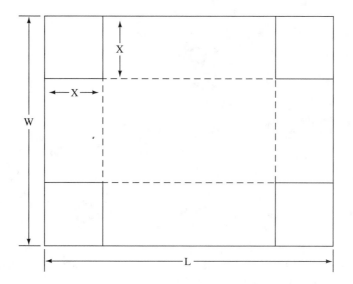

Figure 4.15   The volume of the box is $V = (L - 2X)(W - 2X)X$.

substitute the dimensions of the sheet of paper for L and W: enter $(11 - 2X)(8.5 - 2X)X$. Remember to press ⬚f1 to enter the x (function variable), as you key in the equation. Press ⬚2nd ⬚f2 (<RANGE>), then enter the following: <xMin=0> <xMax=4.25> (The x-value cannot exceed 1/2 of 8.5, the length of the paper, or the volume would be less than 0.) <xScl=1> <yMin=0> <yMax=100> and <yScl=10>. Then press ⬚f5 (<GRAPH>) to display the graph. Press ⬚f4 (<TRACE>) to view the cursor location and the corresponding x and y values at the bottom of the screen. Find the maximum value of y (the volume) by moving the cursor with the arrow keys. At maximum volume the x-value is approximately 1.6 inches. Substituting into the volume equation,

$$(8.5 - 3.2)(11 - 3.2)x = 66.1.$$

The volume is 66.1 cubic inches. The powerful TI-85 offers alternative methods of finding the maximum value of the function. Consult the owner's manual for further information.

Please take time to construct the box. If you compare the box with others in your home, you will find several of similar size. Wonder why?

The TI-85 can also collect data during experiments (Figure 4.16A). The data may be downloaded to a personal computer for large-screen graphs of data (Figure 4.16B).

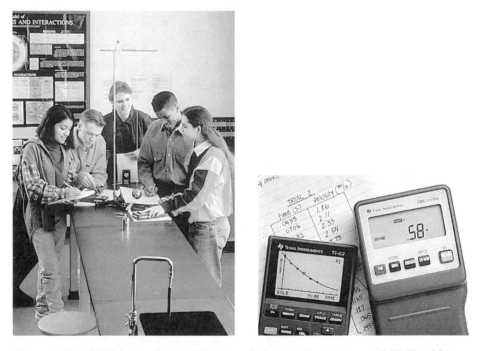

Figure 4.16 (A) Where on the pendulum's arc is the tension the greatest? (B) The CBL system with its own microprocessor is set up with the TI-82 or TI-85 to collect force data from the pendulum experiment. (A, B: Courtesy Texas Instruments, Inc.)

Problems

Section 4.1

1. Visit at least one local electronics store. From its inventory, list the brand name and model number of three calculators that will meet your needs while pursuing an engineering technology education (see Table 4.1). Also, list the price of each and the number of pages in the owner's manual.

2. From the characteristics listed in Table 4.1, list those that your calculator does not possess. Also list any characteristics your calculator possesses that do not appear in Table 4.1.

3. Draw your calculator display, detailing the various symbols shown as you select various functions, i.e., fixed, scientific, or engineering modes; angular modes; and special function modes such as the statistical mode. Also, test the error flag by dividing any number by zero (this operation is undefined, but the technician regards it as infinity [∞]). If the instructor wishes, *TI-85 users should write the first 4 lines of the MODE selection screen.*

Section 4.2

For each of the expressions in Problems 4–10, write the keystroke sequence and the results obtained from the calculations.

4. $4 + 5 - 8 =$ *Example:* 4 ⊞ 5 ⊟ 8 ⊒ 1

5. $1 - 2 - 1 + 7 =$

6. $-9 + 2 + 4 + 3 - 18 =$

7. $(+7) - (+6) - (-4) =$ (See **NOTE:** Why use two signs when one will do? In Example 4.1. Rewrite and simplify the expression before writing the keystroke sequence.)

8. Refer to Example 4.1. If the instructions were to algebraically subtract the last vector (the −1) from the other three, the paper-and-pencil solution would look like this

$$(-3) + (+2) + (+6) - (-1) = +6$$

The minus sign in front of the last parentheses is an operation sign to algebraically subtract, which is to reverse the direction of the last vector (Appendix B). Rewrite and simplify the expression before writing the keystroke sequence.

9. $\dfrac{91 \times 22 \times 35}{12} =$

10. $12 \times 34 + 16 \div 8 =$

11. Diagram Problems 4 and 5 on a number line (see Example 4.1).

Perform the calculations in Problems 12–15. Always write your answer accurate to three significant figures (not decimal places—see Example 4.3 and consult with your instructor to distinguish between significant figure and decimal place accuracies).

12. $52[16 - 4(17 + 3)] =$

Note: The times sign (×) is not written between parentheses, but if there is no sign it is understood to be a multiplication operation (implied multiplication). Be sure to introduce the times sign on the calculator when applicable. The TI-85 recognizes implied multiplication. See *The Equation Operating System* in the *TI-85 Guidebook.*

Example: The keystroke sequence is as follows:

52 [x] [(] 16 [−] 4 [x] [(] 17 [+] 3 [)] [)] [=] − 3328
= −3330—Answer is accurate to three significant figures.

13. $3 \times 5 + 6 - 7 \div 2 =$

14. $\dfrac{2 + 8 + 4 - 3}{5} =$

15. $\dfrac{91 \times 22 \times 68 \times 45}{8} =$

16. An operator uses a vernier caliper to take the following measurements on a part: X_1 = 39.6, X_2 = 39.8, X_3 = 39.6, X_4 = 39.7, X_5 = 39.9. All measurements are in millimeters. Find the average of the measurements and convert the average to inches by using 25.4 mm = 1 in.

17. Explain why the parentheses (or an intermediate "equals" sign) must be used to arrive at the correct answer for the average of the vernier dial caliper in Example 4.3.

18. Use the [f/x] function on your calculator. Enter the number 285.3456 into the calculator by pressing =. By following your operator's manual fix the decimal place to 1, 2, and 3 places. Does your calculator round off or cut off the results? Enter the number 0.00056. Fix the decimal place to 1 and 2 decimal places. Can you lose a small number by unknowingly being in the *fix* mode?

19. TI-85 users press [2nd MODE] and the down arrow key [∇]. Use the right and left arrow keys to move from floating point (Float) to *fix* the decimal point at 0 through 9 places. Perform the operations and answer the questions in Problem 18.

20. $\dfrac{108\ 914}{23.6 \times 28.4 \times 32.5} =$

21. $[(5 - 3)(21 + 17)] \div (46 - 510) =$

*22. $\{[5 + 18(3 - 6)][13 + 2(6 - 5)]\} + 11 =$
Your calculator may not be able to evaluate three levels of parentheses. If it cannot, you must evaluate from the innermost parentheses out to the braces.

23. Refer to Example 4.5. Find the resistance of an aluminum wire (α = 0.004) at −35°C (t_2) if the resistance (R_1) is 20 Ω at 10°C (t_1).

Section 4.3

Solve problems 24–28 and sketch and label the geometric shapes (squares or cubes) each problem represents.

---

*An asterisk (*) indicates a challenging problem—go for it!

24. $(0.002)^2 =$

Example: 0.002 [x^2] [$=$] $0.000\ 004$ in.^{2} (read as 4 millionths square inches)

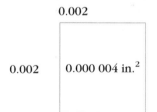

0.002

0.002 | 0.000 004 in.^{2}

Note: A number less than one when raised to a power greater than one results in a *smaller number (i.e., for the area or volume measurement).*

25. $(20\ \text{ft})^2 =$

26. $(0.05\ \text{m})^3 =$

27. $\sqrt{30\ \text{in}^2} =$ (TI-85 users will find that the square root symbol [2nd] [$\sqrt{\ }$] must precede the number.)

28. $\sqrt[3]{36\ \text{cm}^3} =$

*29. $\sqrt{3^2 + 4^2} =$

*30. $\sqrt{6^2 + 8^2} =$

Sketch Problems 31–36 before solving.

31. You must purchase the artificial turf for a football field that measures 120 yd by 160 ft. Find the area to be surfaced in square feet.

*32. In Problem 31, add 20 ft per sideline (there are two sidelines) and recalculate the area in square yards.

33. A sewage treatment plant's aerator ponds are square and cover an area of 6500 ft^2. Find the length of each side.

34. A helicopter landing pad is to be constructed over an area of 40 ft by 70 ft. It is to be 2 ft thick. How many cubic feet of concrete must be ordered?

*35. In Problem 34, solve for the amount of concrete in cubic yards by multiplying by the conversion factor

$$\frac{1^3\ \text{yd}^3}{3^3\ \text{ft}^3} = \frac{1\ \text{yd}^3}{27\ \text{ft}^3}$$

36. If the concrete in Problem 34 is to be 3-ft. thick, what is the volume in cubic feet? Cubic yards?

Use the power key [y^x] or the root key [$\sqrt[x]{}$] to solve Problems 37–44. For the TI-85, use the power key [$\wedge$] and use fractional powers for roots (see Example 4.8).

37. $2^6 =$

38. $-3^5 =$

39. $-3^4 =$ (refer to this answer in back of text)

*40. $33^{-6} =$

---

*An asterisk (*) indicates a challenging problem—go for it!

Example: $33^{-6} = \dfrac{1}{33^6}$

(Refer to Appendix B, Rules of Exponents)

41. $\sqrt[3]{5} =$

42. $\sqrt[8]{14} =$

*43. $95^{1/6} =$

*44. $45^{2/5} =$

Section 4.4

Convert the numbers in Problems 45–49 to scientific notation and then to engineering notation. The power of ten must be evenly divisible by three in order for the power of ten to be substituted with an engineering prefix (i.e., mA $= 10^{-3}$ amperes and microinches μ in. $= 10^{-6}$ in.). Be sure to use your calculator to check your paper-and-pencil solution.

45. 33 000

Example: $33\,000 = 3.3 \times 10^4 \qquad 33 \times 10^3$

$\qquad\qquad\qquad\qquad\quad\uparrow\qquad\qquad\quad\uparrow$

$\qquad\qquad\qquad$ scientific $\qquad$ engineering
$\qquad\qquad\qquad$ notation $\qquad$ notation

46. 820 000 000

47. 0.000 365

48. 0.000 006 65

49. 0.000 000 000 030 40

Convert the numbers in Problems 50–54 to standard (floating point) notation.

50. 563×10^4 *Example:* 5 630 000

51. 966×10^{-7}

52. 22.9×10^3

53. 0.274×10^{-6}

*54. 3020×10^0

For Problems 55–59, use inequality symbols to show whether a number is greater than or less than zero ($3 > 0$ means 3 is greater than 0, $-3 < 0$ means -3 is less than 0).

55. 2.45×10^{-3} *Example:* $2.45 \times 10^{-3} > 0$

56. 445×10^{44}

57. -0.559×10^{83}

58. 0.241×10^{-67}

59. -860×10^{-4}

Solve Problems 60–64 using your calculator.

60. $\dfrac{6 \times 10^{-2} \times 18 \times 10^5}{15 \times 10^{-8}} =$

61. $4^2 \times 6 \times 10^2 =$

---

*An asterisk (*) indicates a challenging problem—go for it!

62. $(1.5 \times 10^{-8})(2.38 \times 10^{-3}) =$

*63. $\dfrac{10^{10}}{5} =$

*64. $3 \times 10^{22} + 8 \times 10^{-12} =$

For Problems 65–69: The mass of the proton is 1.67×10^{-27} kg. Determine how many times greater is the mass of the

65. earth at 5.98×10^{24} kg

66. electron at 9.11×10^{-31} kg

67. sun at 1.99×10^{30} kg

68. neutron at 1.67×10^{-27} kg

Use the reciprocal key $\boxed{1/x}$ to solve Problems 69–75.

69. $\dfrac{1}{8} =$ *Example:* 8 $\boxed{1/x}$ (answer is displayed as 0.125)

70. $\dfrac{1}{3} + \dfrac{1}{2} =$

71. $\dfrac{1}{3 \times 10^7} \times \dfrac{1}{4 \times 10^9} =$

72. $\dfrac{1}{\dfrac{1}{2} + \dfrac{1}{3} - \dfrac{1}{4}} =$

73. $\dfrac{1}{\dfrac{1}{8} \times \dfrac{1}{16}} =$

Hint! For Problems 74 and 75, begin with the powers of ten in the denominator and work towards the top by using the $\boxed{1/x}$ or the $\boxed{x^1}$ keys.

74. $\dfrac{1}{\dfrac{1}{10^3} \times \dfrac{1}{10^4}} =$

75. $\dfrac{1}{\dfrac{1}{2 \times 10^3} + \dfrac{1}{3 \times 10^2} - \dfrac{1}{4 \times 10^3}} =$

Section 4.5

76. On a long trip, you average 62.5 miles per hour. Complete the following table and graph the function on the axes in Figure 4.17.

Hours Traveled	0	1	2	3	4	5	6	7	8
Miles Traveled	0	62.5							

*An asterisk (*) indicates a challenging problem—go for it!

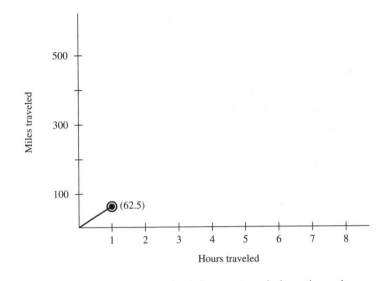

Figure 4.17 A partially completed graph of distance traveled per time, given a constant velocity of 62.5 mph.

*77. Write the keystroke sequence, for the TI-85 calculator, to graph the function $(I = \frac{E}{R})$ in Example 4.15. In this case use a 50 000 ohm resistor $(R= 50\ k\ \Omega)$. *Example:* Press GRAPH. Press [f1] to display the <y(x)=> editor. Enter $y1=x/50E3$. (Press [f1] to enter the x as the function variable.) After entering the rest of the equation, press [2nd] [f2] (<RANGE>), then enter the following: <xMin = 0>, <xMax=10>, <xScl = 1>, <yMin=0>, <yMax=4E−4>, and <yScl = 1E −4>. Then press [f5] (<GRAPH>) to display the graph. Use [f4] (<TRACE>) to approximate the current (I) for different values of voltage (E). Report the current (I) for voltage values (E) of 2, 5, and 7 volts.

*78. Graph the $\sqrt[X]{Y}$ function for $X = 4$ and $Y = 2$ to 10. The first point is plotted on the graph in Figure 4.18. You will find the memory function on the calculator useful. Approximate from the graph the fourth root of 6.5. Check your estimate with your calculator's solution to the fourth root of 6.5.

79. From Example 4.17, how many bits can be stored in a computer that has 128K (kilobytes) of RAM? Note, the graph will not give you the answer.

80. If you have a graphing calculator, graph the function $\sqrt[X]{Y}$ in Problem 78. You may have to make Y, the dependent variable, equal to the constant 4 and vary X from 2 to 10. Use the <TRACE> function key to approximate the fourth root of 6.5.

*81. With a graphing calculator, solve for the maximum volume of a box built from legal-sized paper (8 1/2 in. by 14 in.) See Example 4.18.

---

*An asterisk (*) indicates a challenging problem—go for it!

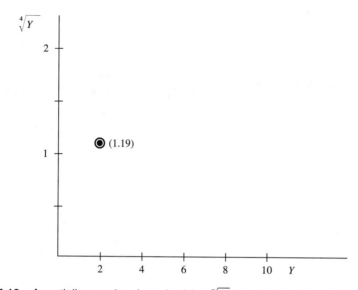

Figure 4.18 A partially completed graph of the $\sqrt[x]{Y}$ function for $X = 4$ and $Y = 2$ to 10.

Selected Readings

Lawrence, David P. *Graphing Calculators: Quick & Easy! The TI-85*. Fairfield, New Jersey: Pencil Point Press, 1993.

Rich, Nelson G. and Gilligan, Lawrence G. *The TI-85 Reference Guide*. Atlanta, Gilmar Publishing, 1993.

Texas Instruments, Incorporated. *TI-85 Guidebook*. Lubbock, Tex., 1992.

Your calculator's owner's manual.

5

Measurement Systems

One of the most important ways business and industrial groups communicate is through systems of units. Most of us have always used some form of the English system. Yet, how many of us can accurately convert cubic feet to cubic yards? Not many of us, even when our buildings and parking lots are usually measured in feet, but the concrete is ordered in yards (actually cubic yards)! To further complicate our business endeavors, the metric system (a global standard) has been used increasingly in countries using the English system. In the U.S. for instance, all future government contracts will use metric units (see Section 5.1)—even the plans for buildings may be dimensioned in metric units. Converting from the English system to metric is not complicated, as this chapter will show you. Your ability to convert and understand these units will aid you in making your company internationally competitive. This chapter will address the

1. fundamental units used in business and industry in the English and metric measurement systems,
2. conversion of units within the English system,
3. conversion of units within the metric system,
4. conversion of units between systems, and
5. use of dimensional analysis in understanding area, volume, and force units.

5.1 The Fundamental Units

Table 5.1 shows the fundamental units. These units establish a base from which all other units are derived. The English units are the same units used in the *U.S. customary system* (USCS) of measurement. It is a close cousin to the cumbersome English system. In this text, what is referred to as the *English system* is actually the U.S. customary system. The USCS was developed from the system used by early American colonists and is still used extensively in the United States.

Table 5.1   The Fundamental Units

	SI (metric) Unit	English Unit
Length	meter	foot
Mass (artifact)	kilogram	slug (the pound—a derived force unit—is most used in the English system)
Time	second	second
Electric current	ampere	ampere
Temperature	Kelvin	Rankine
Luminous intensity	candela	candela
Amount of substance	mole	pound mole

Almost all other countries build products using the *International System of Units* (SI, for Système International, is the internationally accepted abbreviation). This system is a simplification of the metric system, which has been adopted by those other countries as well as by the United States. The U.S. Congress enacted the Metric Conversion Act in 1975, establishing a coordinated national program with time lines for implementation of metrication. In 1988, the Omnibus Trade and Competitiveness Act required that "each Federal agency, by a date certain and to the extent economically feasible by the end of the fiscal year 1992, use the metric system of measurement in its procurements, grants, and other business-related activities, except to the extent that such use is impractical or is likely to cause significant inefficiencies or loss of markets to United States firms, such as when foreign competitors are producing competing products in non-metric units." The United States, to remain competitive and regain a favorable balance of trade, must convert to the metric system.

Even America's schools are affected by the 1975 act, which calls for the metric system to be included in the curriculum of the nation's educational institutions and requires that instructors be properly trained to teach it. Many U.S. industries are implementing the international system. This is a major challenge, because the role of the technician and technologist is implementation.

The **fundamental units** for the SI are defined in the following subsections. The list comprises the seven base units listed in Table 5.1 and two supplementary units. *All other units are derived from the fundamental units.* One of the most important things to note in the briefly defined dimensions that follow is that all except one of the fundamental units are based on physical phenomena reproducible in a laboratory. If you have already taken a high school physics course, this material will be an excellent review. In any case, the material will help you immensely in understanding the college physics course that is part of your engineering technology curriculum. No one physical object, or artifact, is used as a standard except for the mass standard. When you can reproduce a standard in the laboratory, the need for frequent travel to the location of the international primary standard is eliminated.

Length

The meter is defined as the length of the path traveled by light in a vacuum during a time interval of 1/299 792 458 s (see Figure 5.1A). To reproduce the value of the

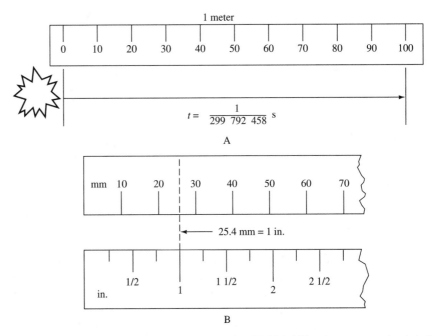

Figure 5.1 (A) The speed of light in a vacuum is 299 792 458 m/s, or approximately 3 × 10^8m/s; (B) The millimeter and inch scales are compared; 1 in. is exactly 25.4 mm.

international standard for length in the laboratory is to measure the wavelength of light emitted by the orange-red line of krypton-86 (krypton is a gas combined with argon in fluorescent lamps). The international meter (compare the millimeter and inch in Figure 5.1B) is now defined as 1 650 763.73 wavelengths of the heated gas.

Mass

The primary standard for the kilogram is a cylinder of platinum-iridium alloy kept by the International Bureau of Weights and Measures in Paris, France. The kilogram is the only fundamental unit still defined by an artifact. An **artifact** is a humanmade object, not as convenient to trace as the laboratory-reproducible standard. The mass standard for the United States is a duplicate of the primary standard (see Figures 5.2A and 5.2B).

Time

A second is the duration of 9 192 631 770 cycles of radiation from the cesium-133 atom. See the "atomic clock" in Figure 5.3.

The SI unit for frequency is the **hertz (Hz),** which is equal to one cycle/second. Standard frequencies, along with correct times, are broadcast from WWV and WWVH. You can access these radio stations with a short-wave radio on frequencies of 2.5, 5, 10, 15, and 20 megahertz (MHz).

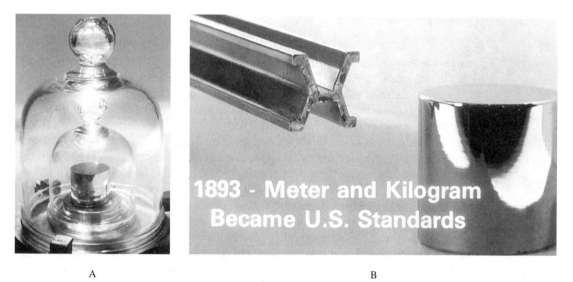

1893 - Meter and Kilogram Became U.S. Standards

A   B

Figure 5.2 (A) The national prototype of the kilogram mass, maintained at the National Institute of Standards and Technology. (B) As early as 1893, the meter and the kilogram became U.S. standards. (Courtesy NIST)

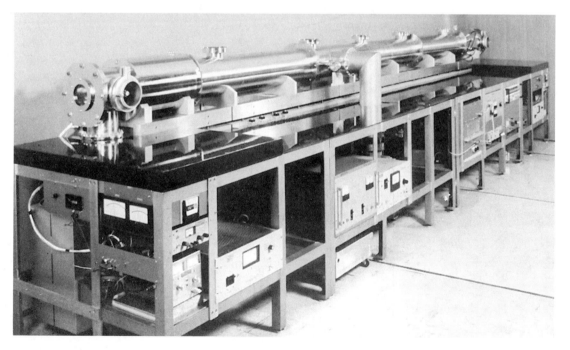

Figure 5.3 The atomic clock, using the resonant frequency of the cesium-133 atom, is maintained in the National Institute of Standards and Technology laboratory in Boulder, Colorado. (Courtesy NIST)

Temperature

The *triple point* of water is that combination of pressure and temperature where water exists as a solid, liquid, and gas at the same time. The temperature at the triple point is an accurate 32.002°F (0.01°C). Standard triple-point cells, maintained in metrology laboratories, are used to calibrate transfer-standard thermometers.

Most of us are familiar with the commonly used Fahrenheit and Celsius temperature scales. Another kind of temperature scale, the **absolute temperature scale** is of tremendous importance to technologists. **Absolute zero** is the theoretical temperature where the molecules of an ideal gas cease to move, and the gas has zero volume. Absolute zero is the Celsius temperature of −273°C. The **Kelvin (K)** temperature scale is set so that its zero is at absolute zero. The size of the Kelvin degree is the same as that of the Celsius degree, so 0 K is equal to −273°C. *Note that the degree sign (°) is not used with the symbol K.*

The absolute temperature scale in the English system is the **Rankine** scale. The size of a Rankin degree is the same as that of the Fahrenheit degree. A diagram of all four temperature scales is shown in Figure 5.4. The following equations are used to convert from Celsius to Fahrenheit, and to each scale's respective absolute temperature scale.

$$°F = 1.8°C + 32°$$
$$°C = (°F − 32°)/1.8$$
$$K = °C + 273°$$
$$°R = °F + 460°$$

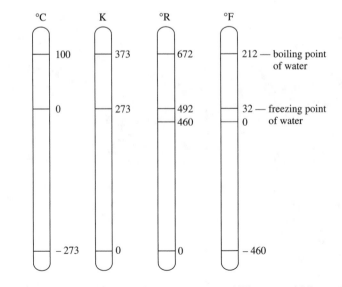

Figure 5.4 The four temperature scales are compared. The two middle scales are absolute temperature scales often used in industry.

Note that to convert from °F to K requires two steps—first the conversion to °C, then the conversion to K. Fortunately, this calculation is seldom necessary. See Problem 5 at the end of this chapter for practice in converting from one temperature scale to another.

It is interesting to realize that the Fahrenheit (English system) scale is more sensitive to temperature change than is the Celsius scale. This fact becomes obvious when you consider the range of the two scales between the freezing and boiling points of water. The freezing and boiling points on the Fahrenheit scale are 32° and 212°, respectively, with a range of 180°. What is the *range* between freezing and boiling points on the Celsius scale—0° to 100°C?

Electric Current

The standard for electric current is the magnetic field existing between two parallel wires. Specifically, it is that current which would produce a repelling force between two wires, one meter long and located one meter apart, of 2×10^{-7} newton·meter (N·m) (see Figure 5.5).

It is important also to know Ohm's law, which states the relationship between the current, voltage, and resistance of an electrical circuit:

$$I = \frac{E}{R}$$

By Ohm's law, one ampere (A) is defined as the current (I) that exists when one volt (V) is applied across one ohm (Ω) of resistance (see Figure 5.6).

Luminous Intensity

The standard for luminous intensity, the **candela,** is established by a source emitting light radiation at 540×10^{12} Hz. One candela of energy will supply the detector with a power of 1/683 watt (W) per steradian (the steradian is defined under "Supplementary Units").

Amount of Substance

The mole is a unit of molecular mass. One **mole** contains the number of atoms in 0.012 kg of carbon-12. It is used mainly by chemical engineering technologists to keep

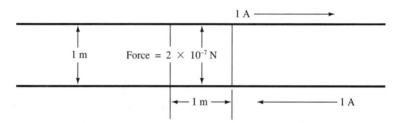

Figure 5.5 The repelling force between two current-carrying conductors is the standard used for electric current.

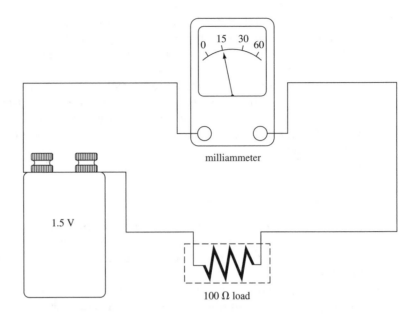

Figure 5.6 A 1.5 V dry cell will supply 0.015 A (15 mA) of current to a 100 Ω load.

$I = E/R = \dfrac{1.5V}{100\Omega} = 0.015A.$

track of the number of atoms or molecules in a sample of matter that is to be blended with another in a batch manufacturing operation. This assures the technologist that the two samples contain a proportional number of molecules. The number of molecules in a mole of any substance is 6.023×10^{23}. See Problem 8 at the end of this chapter to conceptualize how great a number this really is.

Supplementary Units

Two supplementary units are included in the international system—plane and solid angles. The **radian,** the plane angle, with its vertex at the center of a circle, will subtend an arc on that circle equal to the radius. One radian is equal to 57.3 degrees of arc. The **steradian** is a solid angle with its vertex at the center of a sphere subtended by an area of the spherical surface equal to that of a square with sides equal in length to the radius (see Figures 5.7A and 5.7B).

Transfer Standards

Transfer standards are standards traceable to the international (primary) standard, but not at the high accuracy of the primary standard. A **metrology laboratory** provides transfer standards or primary standards for each base unit listed above. All instruments used for precision measurement must be calibrated on a regular basis by using

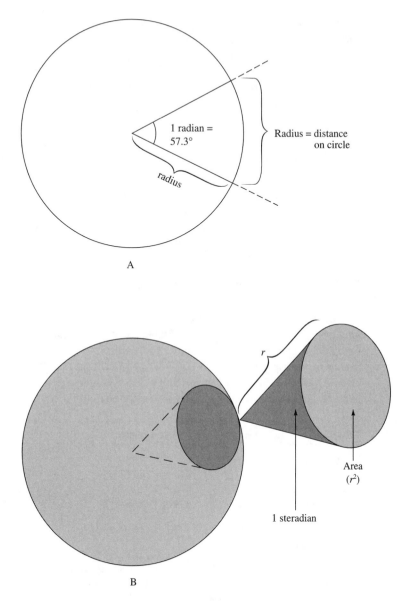

Figure 5.7 The plane and solid angle—the radian and steradian, respectively—are two supplementary units added to the seven base units defined by the international system. (From National Bureau of Standards 304A, courtesy NIST)

transfer standards that ultimately are traceable to the International Bureau of Weights and Measures in Paris. This assurance that a measurement instrument can be related to a primary standard is known as **traceability.** For instance, the standards for the U.S., managed by the **National Institute of Standards and Technology** (NIST), must be traceable to the International Bureau of Weights and Measures in Paris.

The importance of traceability is revealed in *Military Handbook 52A* of 1984. The handbook states

All measuring and test equipment (M&TE) used to assure compliance with the contract must be calibrated by use of standards whose calibration is:

- traceable to National Standards; or
- obtained from independent reproducible standards (derived from accepted values of natural physical constants); or
- traceable to National Standards of a foreign country, which are compared to International or US National Standards.

5.2 Conversion of Units Within the English System

One of the most useful features of the metric system is the simple movement of a decimal point to convert units. Most of us know by now that the kilometer is simply 1000 times the meter. On the other hand, the English system units for length must be converted by multiplying or dividing by some number other than ten. For instance, 3 feet make a yard and 5280 feet make a mile. Easy power-of-ten conversions are found less in the English system than in the metric system. (Refer to Table 5.2.)

Table 5.2 Common English Conversion Factors*

Length	Weight	Time
1 mi[1] = 5280 ft	1 ton = 2000 lb	1 hr = 3600 s
1 yd = 3 ft	1 lb = 16 oz	1 min = 60 s
1 ft = 12 in.	1 slug[5] = 32.2 lb	1 s = 10^3 ms[6]
1 in. = 10^3 thousandths[2]		1 s = 10^6 μs[7]
1 in. = 10^4 tenths[3]		
1 in. = 10^6 μin.[4]		

* See front inside cover, "Table of Equivalents," for complete conversion table.

[1] The statute mile (mi). The nautical mile is 6080 ft.

[2] Thousandth is a precision measurement unit. The thousandth unit means $\frac{1}{1000}$ (0.001) in.

[3] Tenth is often used to indicate $\frac{1}{10}$ of $\frac{1}{1000}$ of an inch, or one ten-thousandth (0.0001) of an inch.

Technicians should be aware of this confusing usage.

[4] The lowercase Greek letter mu (μ) represents one-millionth (0.000 000 1) of an inch, also known as a microinch.

[5] The pound (lb) is a weight, not a mass unit, and is dependent on earth's gravity. This conversion to slug (sl) units is true only on earth, where the acceleration of gravity is 32.2 ft/s².

[6] The millisecond (ms) or 1×10^{-3} (1/1000)s is used frequently in the applied sciences.

[7] The microsecond (μs) or 1×10^{-6} (1/1 000 000)s is also used frequently in the applied sciences.

The following examples of converting English units will employ a technique known as *multiplying by the unit ratio*. Each of the equalities in Table 5.2, when expressed as a ratio, represents a value of one. For example,

$$\frac{12 \text{ in.}}{1 \text{ ft}}, \quad \frac{1 \text{ yd}}{3 \text{ ft}}, \text{ and } \frac{1 \text{ mi}}{5280 \text{ ft}}$$

are ratios that equal unity or one. We refer to these as unit ratios and use them to convert units. The best way to learn the technique is to *work through each step of each example that follows.*

EXAMPLE 5.1

A motor must be mounted on a baseplate measuring 3.76 ft in length. Convert this measurement to inches.

Solution

STEP 1: First, set up for the conversion:

$$\frac{3.76 \text{ ft}}{1} \times \underline{\hspace{2cm}} = \underline{\hspace{2cm}} \text{ in.}$$

Note: The one (1) is placed under the unit value to be converted simply to demarcate the denominator in the term. Similarly, the fraction bar in the second term of the equation demarcates the future placement of the unit ratio to be used to perform the desired conversion.

STEP 2: Next, determine the units that will cancel the feet in the numerator of the first ratio and provide the inches needed in the solution (see Figure 5.8).

STEP 3: Consult Table 5.2 and enter the appropriate numbers into the unit ratio.

$$\frac{3.76 \text{ ft}}{1} \times \frac{12 \text{ in.}}{1 \text{ ft}} = ? \text{ in.}$$

STEP 4: Finally, perform the indicated multiplication.

$$\frac{3.76 \text{ ft}}{1} \times \frac{12 \text{ in.}}{1 \text{ ft}} = 45.1 \text{ in.}$$

You are probably wondering at this point why such a simple conversion requires such a complicated solution. The simple conversion is used to illustrate a powerful method that will allow you to convert even units you are unfamiliar with. Please note that with the method demonstrated above, we required only the conversion table to determine the unit ratio. Had the units been unfamiliar, but we still knew the factors for the unit ratio, we could have completed the conversion. It is important for you to follow each step presented above to be sure of yourself as you continue to learn unit conversions.

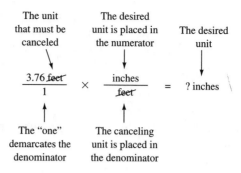

Figure 5.8   A formula showing unit cancellation.

EXAMPLE 5.2

You cannot combine unlike terms or units. The units of one must be converted to the other before they can be added. Find the width of a standard-gauge railroad track (4 ft 8 ½ in.) in feet.

Solution   The solution will be shown for feet. We will follow the same steps established in Example 5.1.

STEP 1: $\dfrac{8.5 \text{ in.}}{1} \times \underline{\hspace{2cm}} = \underline{\hspace{2cm}} \text{ ft}$

STEP 2: $\dfrac{8.5 \text{ in.}}{1} \times \dfrac{\text{ft}}{\text{in.}} = \underline{\hspace{2cm}} \text{ ft}$

STEP 3: $\dfrac{8.5 \text{ in.}}{1} \times \dfrac{1 \text{ ft}}{12 \text{ in.}} = \underline{\hspace{2cm}} \text{ ft}$

STEP 4: $\dfrac{8.5 \text{ in.}}{1} \times \dfrac{1 \text{ ft}}{12 \text{ in.}} = 0.708 \text{ ft}$

Combining the 4 ft and the 0.708 ft, we find the width of the track is 4.708 ft. When you convert to inches your answer is not exact:

$$4.708 \text{ ft} \times 12 \frac{\text{in.}}{\text{ft}} = 56.496 \text{ in.}$$

Using the original information and converting directly to inches: $\dfrac{4 \text{ ft}}{1} \times \dfrac{12 \text{ in.}}{1 \text{ ft}} + 8.5 \text{ in.}$ = 56.5 in. Can you tell why the two answers differ?

EXAMPLE 5.3

A typical professional football player (lineman) weighs 270 lb on earth. What is the lineman's mass in slugs (sl)?

Solution

STEP 1: $\dfrac{270 \text{ lb}}{1} \times \underline{\hspace{2cm}} = \underline{\hspace{2cm}} \text{ sl}$

STEP 2: $\dfrac{270 \text{ lb}}{1} \times \dfrac{\text{sl}}{\text{lb}} = \dfrac{}{} \text{ sl}$

STEP 3: $\dfrac{270 \text{ lb}}{1} \times \dfrac{1 \text{ sl}}{32.2 \text{ lb}} = \dfrac{}{} \text{ sl}$

STEP 4: $\dfrac{270 \text{ lb}}{1} \times \dfrac{1 \text{ sl}}{32.2 \text{ lb}} = 8.39 \text{ sl}$

Up to now we have worked only with fundamental units that can be converted in one step. Consider the following example of the use of intermediate conversion factors.

EXAMPLE 5.4

How many seconds are contained in a period of 70 years?

Solution

STEP 1: $\dfrac{70}{1} \times \dfrac{}{} \times \dfrac{}{} \times \dfrac{}{} = \dfrac{}{} \text{ s}$

Note that three unit ratios are to be constructed: (1) to convert years to days, (2) to convert days to hours, and (3) to convert hours to seconds.

STEP 2: $\dfrac{70 \text{ yr}}{1} \times \dfrac{\text{day}}{\text{yr}} \times \dfrac{\text{hr}}{\text{day}} \times \dfrac{\text{s}}{\text{hr}} = \dfrac{}{} \text{ s}$

STEP 3: $\dfrac{70 \text{ yr}}{1} \times \dfrac{365.25 \text{ day}}{1 \text{ yr}} \times \dfrac{24 \text{ hr}}{1 \text{ day}} \times \dfrac{3600 \text{ s}}{1 \text{ hr}} = \dfrac{}{} \text{ s}$

STEP 4: $\dfrac{70 \text{ yr}}{1} \times \dfrac{365.25 \text{ day}}{1 \text{ yr}} \times \dfrac{24 \text{ hr}}{1 \text{ day}} \times \dfrac{3600 \text{ s}}{1 \text{ hr}} = 2.209 \times 10^9 \text{ s}$

When two or more fundamental units are combined the result is a **derived dimension.** The following example will demonstrate how derived dimensions are converted.

EXAMPLE 5.5

A rocket's velocity is often given in feet per second rather than miles per hour. We can more easily conceptualize speed in miles per hour. Find an easily memorized conversion factor for

$$\frac{\text{mi}}{\text{hr}} \text{ to } \frac{\text{ft}}{\text{s}}.$$

Solution   Use 60 mph and convert to feet per second.

STEP 1: $\dfrac{60 \text{ mi}}{1 \text{ hr}} \times \dfrac{}{} \times \dfrac{}{} = \dfrac{}{} \dfrac{\text{ft}}{\text{s}}$

Note: Two unit conversions are used. One of them will convert miles to feet and the other will convert hours to seconds. Can you confirm the unit ratios in step 2, below? Will they (1) result in the right units canceled; and (2) yield an answer in the appropriate units? Please do not proceed further until you can confirm these ratios.

STEP 2: $\dfrac{60 \text{ mi}}{1 \text{ hr}} \times \dfrac{\text{ft}}{\text{mi}} \times \dfrac{\text{hr}}{\text{s}} = \underline{\hspace{1cm}} \dfrac{\text{ft}}{\text{s}}$

STEP 3: $\dfrac{60 \text{ mi}}{1 \text{ hr}} \times \dfrac{5280 \text{ ft}}{1 \text{ mi}} \times \dfrac{1 \text{ hr}}{3600 \text{ s}} = \underline{\hspace{1cm}} \dfrac{\text{ft}}{\text{s}}$

STEP 4: $\dfrac{60 \text{ mi}}{1 \text{ hr}} \times \dfrac{5280 \text{ ft}}{1 \text{ mi}} \times \dfrac{1 \text{ hr}}{3600 \text{ s}} = 88.0 \dfrac{\text{ft}}{\text{s}}$

Sixty miles per hour equals 88 ft/s. We have a new unit ratio, easily remembered, that converts feet per second to miles per hour.

EXAMPLE 5.6   The velocity of a rocket is 36 667 ft/s when it escapes from earth's orbit. What is the escape velocity in miles/hour?

Solution

STEP 1: $\dfrac{36\,667\ \frac{\text{ft}}{\text{s}}}{1} \times \underline{\hspace{1.5cm}} = \underline{\hspace{1.5cm}} \dfrac{\text{mi}}{\text{hr}}$

A compound fraction is formed. Note how effective the one (1) placed in the denominator is for demarcating the compound fraction necessary for the conversion.

STEP 2: $\dfrac{36\,667\ \frac{\text{ft}}{\text{s}}}{1} \times \dfrac{\frac{\text{mi}}{\text{hr}}}{\frac{\text{ft}}{\text{s}}} = \underline{\hspace{1cm}} \dfrac{\text{mi}}{\text{hr}}$

STEP 3: $\dfrac{36\,667\ \frac{\text{ft}}{\text{s}}}{1} \times \dfrac{60\ \frac{\text{mi}}{\text{hr}}}{88\ \frac{\text{ft}}{\text{s}}} = \underline{\hspace{1cm}} \dfrac{\text{mi}}{\text{hr}}$

STEP 4: $\dfrac{36\,667\ \frac{\text{ft}}{\text{s}}}{1} \times \dfrac{60\ \frac{\text{mi}}{\text{hr}}}{88\ \frac{\text{ft}}{\text{s}}} = 25\,000 \dfrac{\text{mi}}{\text{hr}}$

We have just converted the rocket's velocity from feet/second to miles/hour. You can now add this unit ratio

$$60 \frac{\text{mi}}{\text{hr}} = 88 \frac{\text{ft}}{\text{s}}$$

to a conversion table for derived units.Unit conversion often involves converting from one physical dimension to another by using physical conversion factors such as weight density (weight/unit volume). The following example demonstrates a conversion from gallons (volume) to pounds (weight).

EXAMPLE 5.7

The F4-E Phantom, a U.S. fighter/bomber aircraft, carries 370 gal of fuel in each wing-tank. Pilots need to know the weight of the fuel at takeoff. Convert the fuel in one wing-tank to pounds. The weight density of the fuel is $6.5 \frac{lb}{gal}$, which means that each gallon of jet fuel weighs 6.5 lb (when no denominator number is written it is understood to be 1, or in this case, $\frac{6.5 \, lb}{1 \, gal}$).

Solution   First we must decide how the density term is to be used. Set up Step 1 and then proceed to Step 2. The third step will become evident at this point.

$$\textbf{STEP 1: } \frac{370 \text{ gal}}{1} \times \underline{\hspace{2cm}} = \underline{\hspace{1.5cm}} \text{ lb}$$

$$\textbf{STEP 2: } \frac{370 \text{ gal}}{1} \times \frac{\text{lb}}{\text{gal}} = \underline{\hspace{1.5cm}} \text{ lb}$$

The density ratio's use becomes obvious at this point. It gives us a unit ratio that may be used to convert gallons to pounds.

$$\textbf{STEP 3: } \frac{370 \text{ gal}}{1} \times \frac{6.5 \text{ lb}}{1 \text{ gal}} = \underline{\hspace{1.5cm}} \text{ lb}$$

$$\textbf{STEP 4: } \frac{370 \text{ gal}}{1} \times \frac{6.5 \text{ lb}}{1 \text{ gal}} = 2410 \text{ lb (rounded to 3-place accuracy)}$$

Over 2400 pounds on one wing! It is important for a pilot to know the weight of this large amount of fuel.

5.3   Conversion of Units Within the International System

We will leave the English system to consider metric conversion. After completing this section, you will perhaps wonder why many industries in the United States have been reluctant to convert to the metric system. It is much easier to move a decimal point than to have to multiply and divide by number bases other than ten, such as 12 (in./ft), 3 (ft/yd), and 5280 (ft/mi). Advantages other than the decimal (ten) base for conversion include (1) fewer and less confusing basic units; and (2) the use of prefixes keyed to powers of ten.

But the main incentive for U.S. metrication is to regain a greater share of world markets. The rest of the world is on the metric system and wants products with metric dimensions.

The 1984 Olympics in Los Angeles used only metric units, and Americans accepted this use of metrics. The liquor industry has already reduced costs by standard-

Table 5.3 Most Used Engineering Prefixes (May Be Used with All SI Units)

Number	Power of Ten	Prefix	Symbol
1 000 000 000	10^9	giga-*	G
1 000 000	10^6	mega-	M
1 000	10^3	kilo-	k**
1	10^0	none	none
0.001	10^{-3}	milli-	m
0.000 001	10^{-6}	micro-	μ
0.000 000 001	10^{-9}	nano-	n
0.000 000 000 001	10^{-12}	pico-	p

*Pronounced jiga.
**Lower case k used for kilo (upper case K used for Kelvin)

izing bottle sizes in metric sizes. Today, only two metric bottle heights are used, so shelves can be standardized. The complete implementation of the metric system can—and should—be accomplished soon for those companies competing in world markets.

In the early 1970s the auto industry recognized the need to manufacture cars for world markets. The international system had to be adopted to most effectively serve these markets. Implementation has been arduous and slow, however. Today's American- and British-made automobiles are a mixture of English and metric sizes.

You have already become familiar with engineering prefixes in the English system, such as microinch and millisecond. Table 5.3 is not complete, but lists the most commonly used metric prefixes. *It is recommended that this table be memorized. All of the prefixes are engineering prefixes that represent powers of ten evenly divisible by three.* For instance, the prefix centi- is not shown on the table. The centimeter is $\frac{1}{100}$ meter or 10^{-2} m. The centimeter may be used to describe area and volume. An example is the use of cubic centimeters (cc) to describe the displacement of a motorcycle engine. Engineering prefixes can be converted on the calculator automatically by pressing ENG or FSE keys after a number is entered.

EXAMPLE 5.8

Convert 100 meters to kilometers and to millimeters.

Solution   First, convert to kilometers by

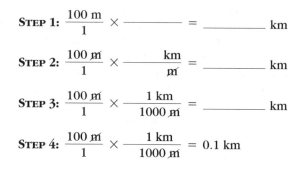

$$\textsc{Step 1:} \quad \frac{100 \text{ m}}{1} \times \underline{\qquad\qquad} = \underline{\qquad\qquad} \text{ km}$$

$$\textsc{Step 2:} \quad \frac{100 \text{ m}}{1} \times \frac{\text{km}}{\text{m}} = \underline{\qquad\qquad} \text{ km}$$

$$\textsc{Step 3:} \quad \frac{100 \text{ m}}{1} \times \frac{1 \text{ km}}{1000 \text{ m}} = \underline{\qquad\qquad} \text{ km}$$

$$\textsc{Step 4:} \quad \frac{100 \text{ m}}{1} \times \frac{1 \text{ km}}{1000 \text{ m}} = 0.1 \text{ km}$$

Next, convert to millimeters by

STEP 1: $\dfrac{100 \text{ m}}{1} \times \text{———} =$ ———— mm

STEP 2: $\dfrac{100 \text{ m}}{1} \times \dfrac{\text{mm}}{\text{m}} =$ ———— mm

STEP 3: $\dfrac{100 \text{ m}}{1} \times \dfrac{1 \text{ mm}}{1 \times 10^{-3} \text{ m}} =$ ———— mm

STEP 4: $\dfrac{100 \text{ m}}{1} \times \dfrac{1 \text{ mm}}{1 \times 10^{-3} \text{ m}} = 1 \times 10^2 \times 1 \times 10^3$

$$= 1 \times 10^5 \text{ mm} = 100\ 000 \text{ mm}$$

Because there are 1000 mm in 1 m, Step 4 could be written as

STEP 4: $\dfrac{100 \text{ m}}{1} \times \dfrac{1000 \text{ mm}}{1 \text{ m}} = 100 \times 1000 = 100\ 000 \text{ mm}$

Can you now see that 0.1 km is equivalent to 10^5 mm?

EXAMPLE 5.9   Convert 11 300 grams (g) of mass to kilograms of mass.

Solution

STEP 1: $\dfrac{11\ 300 \text{ g}}{1} \times \text{———} =$ ———— kg

STEP 2: $\dfrac{11\ 300 \text{ g}}{1} \times \dfrac{\text{kg}}{\text{g}} =$ ———— kg

STEP 3: $\dfrac{11\ 300 \text{ g}}{1} \times \dfrac{1 \text{ kg}}{1000 \text{ g}} =$ ———— kg

STEP 4: $\dfrac{11\ 300 \text{ g}}{1} \times \dfrac{1 \text{ kg}}{1000 \text{ g}} = 11.3 \text{ kg}$

EXAMPLE 5.10   Human hair grows about 2×10^{-5} meters per hour. To better visualize this speed, convert it to meters per day, centimeters per day, and millimeters per day.

Solution   For m/day,

STEP 1: $\dfrac{2 \times 10^{-5} \text{ m}}{1 \text{ hr}} \times \text{———} = \text{————} \dfrac{\text{m}}{\text{day}}$

STEP 2: $\dfrac{2 \times 10^{-5} \text{ m}}{1 \text{ hr}} \times \dfrac{\text{hr}}{\text{day}} = \text{————} \dfrac{\text{m}}{\text{day}}$

STEP 3: $\dfrac{2 \times 10^{-5} \text{ m}}{1 \text{ hr}} \times \dfrac{24 \text{ hr}}{1 \text{ day}} = \underline{\hspace{2cm}} \dfrac{\text{m}}{\text{day}}$

STEP 4: $\dfrac{2 \times 10^{-5} \text{ m}}{1 \text{ hr}} \times \dfrac{24 \text{ hr}}{1 \text{ day}} = 4.8 \times 10^{-4} \dfrac{\text{m}}{\text{day}} = 0.000\,48 \dfrac{\text{m}}{\text{day}}$

or, 4.8×10^{-4} m/day $= 480 \times 10^{-6}$ m/day (you may wish to use the $\boxed{\text{ENG}}$ key on your calculator to confirm this), and the 10^{-6} can be replaced by the prefix micro (μ). The result is 480 micrometers/hour, or 480 μm/hr. The μm was formerly known in the U.S. as the micron.

Solution For cm/day,

STEP 1: $\dfrac{4.8 \times 10^{-4} \text{ m}}{1 \text{ day}} \times \underline{\hspace{2cm}} = \underline{\hspace{2cm}} \dfrac{\text{cm}}{\text{day}}$

STEP 2: $\dfrac{4.8 \times 10^{-4} \text{ m}}{1 \text{ day}} \times \dfrac{\text{cm}}{\text{m}} = \underline{\hspace{2cm}} \dfrac{\text{cm}}{\text{day}}$

STEP 3: $\dfrac{4.8 \times 10^{-4} \text{ m}}{1 \text{ day}} \times \dfrac{1 \text{ cm}}{10^{-2} \text{ m}} = \underline{\hspace{2cm}} \dfrac{\text{cm}}{\text{day}}$

STEP 4: $\dfrac{4.8 \times 10^{-4} \text{ m}}{1 \text{ day}} \times \dfrac{1 \text{ cm}}{10^{-2} \text{ m}} = 4.8 \times 10^{-2} \dfrac{\text{cm}}{\text{day}}$ or,

$48 \times 10^{-3} \dfrac{\text{cm}}{\text{day}}$ (engineering notation)

Solution For mm/day, it is easy to convert to a different prefix in the SI system, because it will always change by some power of ten. In this case, changing to the prefix milli- (10^{-3} units) must be compensated for by multiplying the number by 10. (Visualize that a millimeter is 10 times smaller than a centimeter.) Therefore,

$4.8 \times 10^{-2} \dfrac{\text{cm}}{\text{day}} \times 10 \dfrac{\text{mm}}{\text{cm}} = 0.48 \dfrac{\text{mm}}{\text{day}}$, or about one-half millimeter per day.

The international system is not "cast in concrete," but is constantly undergoing improvement. The General Conference on Weights and Measures includes representatives from all over the world. As standards are set or agreed on, they are documented and coordinated worldwide.

Much of what the General Conference does is to standardize the conventions used to communicate the numbers, words, prefixes, and symbols of the international system. Refer to the list below and use the accompanying examples to learn the communication standards of the international system.

- Never use common fractions: 0.333 kg, not 1/3 kg
- Never mix units: 5.500 m, not 5 m 500 mm
- Put a zero before a leading decimal point (most calculators automatically do this): 0.787 mm, not .787 mm

- Never use periods or an "s" at the end
  of a unit: 12 kg, not 12 kgs
- Avoid the use of special names for
  multiples of SI units: micrometer (μm) not micron

During your technical career you will be expected to know and use accepted conventions. The correct use of such conventions allows clear and efficient technology transfer between individuals, companies, and countries.

5.4 Conversion of Units Between Systems

The most important part of all the previous material in this chapter is to allow you to easily and accurately convert units. Someday, you will probably be asked to train others or at least to explain unit conversion to others. Complete conversion to metric will rely heavily on the technician and technologist, the implementors of business and industry.

Conversion between units is based on the unit ratio that has already been used to convert units within systems. With a properly composed unit ratio you may be sure that your conversion will yield the units you desire and that the numbers (amounts) the new units are equal to will be computed correctly. Consequently, the examples that follow adhere to the four-step solution process described earlier. In the solution process you

1. established the format for multiplying a ratio (the unit ratio selected in the next step) to obtain the desired units,
2. inserted the desired units so that the units we wish to get rid of cancel, and the units we wish to change to are in the correct location (numerator or denominator),
3. inserted the values of the numbers of the new units so a unit ratio is formed, and
4. performed the calculations that yielded the correct answer in the desired, new units.

It must be emphasized just how important it is to follow each of the steps in order and not try to do two steps at once. Confidence in unit conversion will be possible only if the steps are performed in the proper sequence and always performed one step at a time. Table 5.4 will be used in the examples that follow.

Table 5.4 Common English to Metric Conversion Factors*

Length	Weight	Time
1 mi (5280 ft) = 1.61 km	1 lb† = ˄0.454 kg	1 hr = 3600 s
1 yd = ˄0.9144 m‡	1 slug† = 14.6 kg	1 min = 60 s
1 ft = ˄0.3048 m		1 s = 10^3 ms
1 in. = ˄25.4 mm		1 s = 10^6 μs

*See inside front cover for complete conversion table.
†The pound (lb) is a unit of force and is equal to 0.454 kg only on the earth's surface. The slug, a unit of mass like the kilogram, is everywhere equivalent to 14.6 kg.
‡ ˄ = exact.

EXAMPLE
5.11

The Olympic running track is 400 m. Convert the length of the oval to yards (the units used in the United States).

Solution Using the four-step approach listed above,

STEP 1: $\dfrac{400 \text{ m}}{1} \times \underline{\hspace{3cm}} = \underline{\hspace{2cm}} \text{ yd}$

STEP 2: $\dfrac{400 \text{ m}}{1} \times \dfrac{\text{yd}}{\text{m}} = \underline{\hspace{2cm}} \text{ yd}$

STEP 3: $\dfrac{400 \text{ m}}{1} \times \dfrac{1 \text{ yd}}{0.9144 \text{ m}} = \underline{\hspace{1.5cm}} \text{ yd}$

STEP 4: $\dfrac{400 \text{ m}}{1} \times \dfrac{1 \text{ yd}}{0.9144 \text{ m}} = 437 \text{ yd}$

The 400-m track is very close to the length of the quarter-mile (440-yard) track used in the United States.

EXAMPLE
5.12

Great Britain's XJ6 Jaguar automobile is based on English dimensions. Convert the following dimensions to metric: (1) wheelbase = 113.0 in.; (2) curb weight = 3903 lb; (3) fuel tank capacity = 23.2 U.S. gal.

Solution 1

$$\frac{113.0 \text{ in}}{1} \times \frac{25.4 \text{ mm}}{1 \text{ in}} \times \frac{1 \text{ m}}{1000 \text{ mm}} = 2.87 \text{ m (wheelbase)}$$

Confirm the four-step solution although all four solution steps are not shown.

Solution 2

$$\frac{3903 \text{ lb}}{1} \times \frac{0.454 \text{ kg}}{1 \text{ lb}} = 1.77 \times 10^3 \text{ kg (curb weight)}$$

Note: The **newton** is a more acceptable unit of force or weight. Force units are discussed in Section 5.5.

Solution 3   The **liter** is the SI volume measurement. From the conversion table (inside front cover), use 3.79 L/1 gal as the unity conversion factor.

$$\frac{23.2 \text{ gal}}{1} \times \frac{3.79 \text{ L}}{1 \text{ gal}} = 87.9 \text{ L (fuel tank capacity)}$$

It is useful to note that volume in liters is approximately four times the volume measure in U.S. gallons.

EXAMPLE 5.13   The speedometer on the XJ6 ends at 150 miles per hour (mph). Convert this speed to kilometers per hour.

Solution

$$\frac{150 \text{ mi}}{1 \text{ hr}} \times \frac{1.61 \text{ km}}{1 \text{ mi}} = 242 \frac{\text{km}}{\text{hr}}$$

EXAMPLE 5.14   **Torque** is defined as a force acting at some radius (moment arm) or, expressed as an equation: torque = force × moment arm. The units for torque are given as the foot · pound (ft · lb) or inch·pound (in. · lb) in the English system and as the newton · meter (N · m) in the international system. One newton equals 0.225 lb of force. Find the torque in Figure 5.9 in ft · lb and in N · m if the wrench handle is 18 in. long and the applied force is 20 lb.

Solution   First find the torque in inch · pounds by 20 lb × 18 in. = 360 lb · in. Then, convert to ft · lb by

$$\frac{360 \text{ lb} \cdot \text{in.}}{1} \times \frac{1 \text{ ft}}{12 \text{ in.}} = 30 \text{ ft} \cdot \text{lb}$$

Finally, convert to N · m:

$$\frac{30 \text{ ft} \cdot \text{lb}}{1} \times \frac{1 \text{ N}}{0.225 \text{ lb}} \times \frac{0.305 \text{ m}}{1 \text{ ft}} = 40.7 \text{ N} \cdot \text{m}$$

This last example indicates how effective the use of the unit ratio(s) can be. It is now possible for you to convert units you have never worked with to other units by simply following the procedures in this chapter. Another principle introduced in the example is the substitution of units for physical dimensions (in this case, substituting lb · in. for force times distance of moment arm). This leads us into the next section—dimensional analysis.

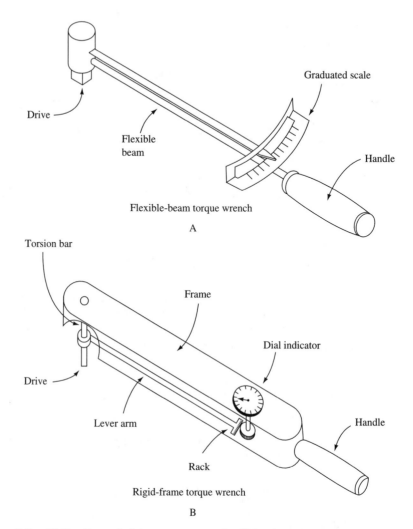

Figure 5.9 (A) The "beam" of the torque wrench will flex in proportion to the torque applied. (B) The rigid frame torque wrench has a dial indicator readout. A third type of torque wrench may be preset to a desired torque, and upon reaching the desired torque a "click" is heard.

5.5 The Use of Dimensional Analysis

To understand dimensional analysis you must be able to use some basic rules of algebra. These rules, mainly involving ratios or fractions, enable you to verify that an equation is correct. In other words, if the units on both sides of the equation do not agree then there must be an error on one side or the other. To illustrate dimensional analysis, we will first use the dimensions of area and volume.

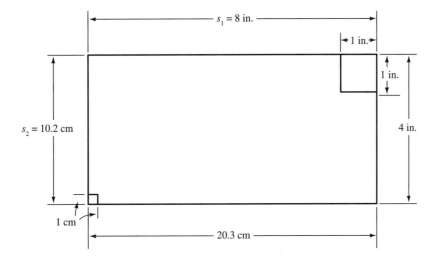

Figure 5.10   The rectangle's area is $s_1 \times s_2 = 4 \times 8$ in.^{2} or 10.2×20.3 cm^2.

Area, as was discussed in Chapter 4, pertains to a two-dimensional figure. The length and width of the figure, when multiplied, equal the area of the figure. See Figure 5.10, where both metric and English units are used to describe the length of the sides.

By using some algebra with dimensional analysis we can define area as

Area (A) = length $\times$ width = $s \times s = s^2$

Note: The symbol for the basic physical dimension of length is L, and for the linear displacement of a particle is s. This text will use the symbol s for both length and displacement.

$A = s \times s = s^2 =$ in. $\times$ in. (in.^{2}) or cm $\times$ cm (cm^2)

To summarize, the steps shown illustrate the power of dimensional analysis. Dimensional analysis yields units that are dimensionally correct for all physical quantities. As you become better at managing units through dimensional analysis concepts, you will have more confidence in handling the applied science courses in your curriculum.

EXAMPLE
5.15

Use dimensional analysis to solve for the area of the rectangle in Figure 5.10.

Solution

$A = s_1 \times s_2$ (representing the length [s_1] and the width [s_2]) = 4 in. $\times$ 8 in. = 32 in.^{2}

$= s_1 \times s_2 = [4$ in. $\times$ 2.54 cm/in.$] \times [8$ in. $\times$ 2.54 cm/in.$] = 207$ cm^2

Can we convert directly from square inches to square centimeters? Yes, but only if we obey the concepts of dimensional analysis. One of the most important

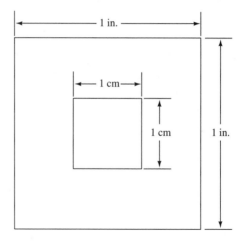

Figure 5.11 One in.^{2} contains 6.45 cm^2.

understandings is that *when a unit is raised to a power, the associated number must be raised to the same power.* In Figure 5.11, observe that the result of multiplying the 4 times the 8 in. could not be equivalent to a value in centimeters by using a conversion factor of 2.54. The conversion factor must be squared. Dimensionally,

$$\frac{32 \text{ in.}^2}{1} \times \frac{2.54^2 \text{ cm}^2}{1^2 \text{ in.}^2} = 206 \text{ cm}^2 \qquad \text{(a slightly different answer due to rounding differences)}$$

Note that the 1 when squared still equals 1. We could write a new conversion factor as 1 in.^{2} = 2.54^2, or 6.45 cm^2.

EXAMPLE 5.16

Show by dimensional analysis that volume (V) Is equal to s^3 (Figure 5.12).

Solution

$$V = s_1 \times s_2 \times s_3 = s \times s \times s = s^3.$$

EXAMPLE 5.17

Given the length, width, and height of the cube in Figure 5.12 as 1 in. ($V = 1$ in.^{3}) solve for the volume in cm^3.

Solution

$$\frac{1 \text{ in.}^3}{1} \times \frac{2.54^3 \text{ cm}^3}{1^3 \text{ in.}^3} = 16.4 \text{ cm}^3$$

Had the answer not yielded cubic centimeters we would know that it must be incorrect. By using dimensional analysis we knew the conversion factor of 2.54 had to

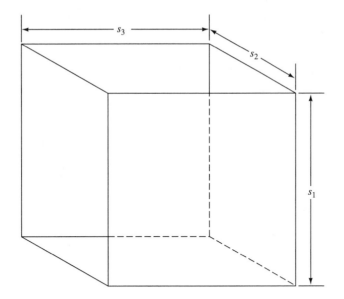

Figure 5.12 Use the cube to visualize a volume's three dimensions.

be cubed also. Dimensional analysis prevents embarrassing mistakes when converting areas and volumes.

EXAMPLE 5.18 The XJ6 Jaguar (Example 5.12) 6-cylinder engine has cylinder dimensions as follows: bore = 3.583 in., stroke = 3.622 in. Find the total engine displacement in cubic inches (cu in.), cubic centimeters (cc [cm^3]), and liters (L).

Solution First sketch and dimension one cylinder (Figure 5.13). Using dimensional analysis and by visualizing the sketch we can write

$$V = A \times s = s^3$$

for each of the 6 cylinders. The area of the top of the cylinder is defined as

$$A = \pi r^2 \text{ or } \frac{\pi d^2}{4}$$

Because we have diameter, use the $\frac{\pi d^2}{4}$. The equation becomes

$$V = A \times s = \frac{\pi d^2}{4} \times s = \frac{\pi (3.583)^2}{4} \times 3.622 = 36.5 \text{ in.}^3$$

Therefore 36.5 cu in. represents the volume of each of the 6 cylinders. Total displacement is 6×36.5 in.^{3} = 219 in.^{3}

Next, convert the displacement to cubic centimeters by

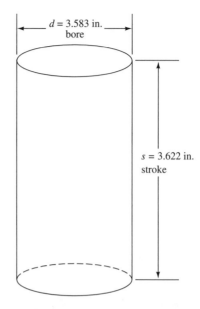

Figure 5.13 To find total engine displacement, solve for the volume of one cylinder and multiply by six.

$$\frac{219 \text{ in.}^3}{1} \times \frac{2.54^3 \text{ cm}^3}{1^3 \text{ in.}^3} = 3590 \text{ cc}$$

An easy conversion factor to remember is that there are 1000 cc/L. By moving the decimal point three places, we can make the final conversion, to find the XJ6 engine displacement to be 3.6 L.

Dimensional analysis can help us understand English and SI force units. To find the dimensions for force (F) we must use **Newton's second law of motion.** In words, his law states: "A body of mass (m) which is acted upon by an unbalanced force is given an acceleration (a) in the direction of the force which is proportional to the force acting and inversely proportional to the mass of the object." The equation that represents his law is much shorter and is stated as

$$a = \frac{F}{m}$$

and, by multiplying both sides by m, we get the more familiar statement $F = ma$. To understand the difference between force and mass, you must first understand acceleration (a speeding up or slowing down—Figure 5.14) and inertia (Newton's first law). This understanding will be acquired in the physics laboratory. For now, we will simply define acceleration as a speed change, with dimensions of length (s) per time squared (t^2) and use the second law to solve for the correct English and SI units of force. We will use the results to derive a conversion factor for pounds to newtons (lb to N).

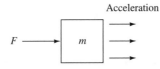

Figure 5.14 A force (F) on a mass (m) causes an acceleration (a)—a speeding up. Slowing down is reported as negative acceleration.

EXAMPLE 5.19

Use the equation $F = ma$ to solve for the basic units for the pound and newton. Evaluate the units to find a conversion factor to convert pounds to newtons.

Solution

$$F \text{ in pounds} = ma = \text{slugs (sl)} \times \frac{\text{ft}}{\text{s}^2} = \frac{\text{sl} \cdot \text{ft}}{\text{s}^2}$$

$$F \text{ in newtons} = ma = \text{kilograms (kg)} \times \frac{\text{m}}{\text{s}^2} = \frac{\text{kg} \cdot \text{m}}{\text{s}^2}$$

To convert pounds to newtons:

$$\frac{\text{lb}}{1} \times \frac{\text{N}}{\text{lb}} = \frac{\dfrac{\text{sl} \cdot \text{ft}}{\text{s}^2}}{\dfrac{1}{1}} \times \frac{\dfrac{\text{kg} \cdot \text{m}}{\text{s}^2}}{\dfrac{\text{sl} \cdot \text{ft}}{\text{s}^2}} = \frac{\text{kg} \cdot \text{m}}{\text{s}^2} = \text{newtons}$$

The unit ratio, represented by the second complex fraction, is evaluated by finding how many $\frac{\text{kg} \cdot \text{m}}{\text{s}^2}$ is in one $\frac{\text{sl} \cdot \text{ft}}{\text{s}^2}$. In other words,

$$\frac{1 \text{ sl} \cdot \text{ft}}{\text{s}^2} \times \underline{\hspace{2cm}} = \underline{\hspace{2cm}} \frac{\text{kg} \cdot \text{m}}{\text{s}^2}$$

and the ratios are

$$\frac{1 \text{ sl} \cdot \text{ft}}{\text{s}^2} \times \frac{\text{kg}}{\text{sl}} \times \frac{\text{m}}{\text{ft}} = \underline{\hspace{2cm}} \frac{\text{kg} \cdot \text{m}}{\text{s}^2}$$

From Table 5.4,

$$\frac{1 \text{ sl} \cdot \text{ft}}{\text{s}^2} \times \frac{14.6 \text{ kg}}{1 \text{ sl}} \times \frac{0.305 \text{ m}}{1 \text{ ft}} = \underline{\hspace{2cm}} \frac{\text{kg} \cdot \text{m}}{\text{s}^2}$$

and, after multiplying, we find there are $4.45 \text{ kg} \cdot \text{m/s}^2$ in one $\text{sl} \cdot \text{ft/s}^2$, which is the same as stating that there are 4.45 N in one lb.

This last example will be difficult for you to fully comprehend without some physical concepts that will be explained later. The reason it was shown here was to heighten your awareness of (1) how complicated units and dimensions can be; and (2) how they may be understood more completely by using two tools—the unit ratio and dimensional analysis.

5.6 Precision of the Inch and Millimeter Scales

The concepts of accuracy and precision will be discussed in Chapter 7, "The Laboratory." But we can begin to think about precision by considering the differences in a common metric-to-English conversion. The English, or customary, system is often referred to as the **inch · pound system,** and the inch is used to establish product dimensions. The **millimeter** is the SI unit used when converting from inches to metric units. For instance, steel wire gage* number 2 is 0.262 in. in diameter and the millimeter equivalent is 6.65 mm (see Problem 29). *Note that the millimeter equivalent is accurate to two decimal places and not to three decimal places as is the inch measurement.* This is because the millimeter is a smaller unit than the inch.

Because there are 25.4 mm in 1 in., the millimeter is smaller and more precise. (Compare the millimeter and inch scales in Figure 5.1B.) **Precision** for measuring instruments has to do with the distance between the marks on a scale. Therefore, because the millimeter scale is more precise, when inches are converted to millimeters the decimal place accuracy will not be the same. For instance, consider measuring an object to a thousandth of an inch. The uncertainty (the opposite of precision) is ±0.0005, or up to five ten-thousandths of an inch. Using 25.4 mm per inch

$$\frac{0.0005 \text{ in.}}{1} \times \frac{25.4 \text{ mm}}{1 \text{ in.}} = 0.0127 \text{ mm}$$

Rounded off, our uncertainty in mm is 0.01 mm.

EXAMPLE 5.20

The target dimension of the length of the shaft for a fractional horsepower motor is 3.025 inches. Convert this dimension to millimeters. Round off to the appropriate decimal place.

Solution

$$\frac{3.025 \text{ in.}}{1} \times \frac{25.4 \text{ mm}}{1 \text{ in.}} = 76.835 \text{ mm}$$

Due to the increased precision of the millimeter scale, the target measurement in mm is 76.84 mm.

A good rule of thumb is to convert inches at a three-decimal-place accuracy to millimeters at only a two-decimal-place accuracy. In fact, a gage or caliper graduated in divisions of 0.02 mm is comparable to one graduated in divisions of 0.001 in.

---

*In metrology, it is common practice to use *gage* for dimensional measurements and *gauge* for all other quantities measured, such as force or pressure.

The important idea to retain from this discussion is that differing units are at differing levels of precision and accuracy. When converting units in the laboratory or on the job, this difference must be considered.

Many modern calculators have built-in conversion factors. For instance, the TI-85 has conversions which are accessed from the [CONV] menu. To access the menu press [2nd] [CONV] . Select the type of conversion by pressing one of the [F-1 through 5] keys. After inputting your measurement data, select the *from* unit. Then press the *to* unit and [ENTER] to complete the calculation. In the previous example the key presses are

[2nd] [CONV] [f1] 3.025 [f4] [f1] [ENTER]

Calculator conversions were not pointed out earlier in this chapter because learning the unit-ratio conversion method will definitely help you to better understand the importance of units, as well as allowing you the freedom to convert units without benefit of a calculator. I hope that if you use your calculator conversions in solving the following problems that you will check the calculator's work with your understanding of unit conversion by using the unit-ratio method.

Applying the concepts in this chapter will help you to become more independent in your technical career. A firm understanding of the use and importance of units is vital to furthering your career goals in technology.

Problems

Section 5.1

1. For the following listed units, classify each as English (E), Système International (SI), or both (B); also classify each as a fundamental unit (f) or a derived unit (d).

Unit		Classification
a.	candela	B, f
b.	meter	
c.	horsepower	
d.	second	
e.	miles/hour	
f.	foot	
g.	square yard	
h.	liter	
i.	ampere	
j.	Kelvin	
k.	ohm/meter	
l.	pounds (see Example 5.19)	
m.	gallons	
n.	Celsius	
o.	kilogram	
p.	cm^3	
q.	mole	

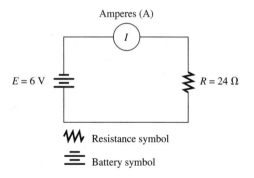

Figure 5.15 A schematic diagram to be used to solve for the current (I) in Problem 7.

2. Metrology is defined in most dictionaries. Write the definition.

3. Define an artifact as related to the establishment of a standard. Which fundamental unit is still based on an artifact? Explain how the meter's standard is based on something other than an artifact.

4. What is a transfer standard? How can one be used in a manufacturing facility?

5. Perform the following temperature conversions. Use the equations °F = 1.8°C + 32, and °C = (°F − 32)/1.8 for items a through c:

   a. 48°F to °C

   b. −398°F to °C

   *c. −40°C to °F

   d. −86°C to K (no degree symbol is used with the absolute zero Kelvin scale)

   e. 1024°C to K

6. Which is the more sensitive temperature scale, or the scale that changes most with the same change in temperature—the Fahrenheit or Celsius scale? Consider both scales to be indexed in one-degree increments.

7. For the circuit of Figure 5.15, solve for the current (I). Use Ohm's law:

$$I = \frac{E}{R}$$

*8. In a 22.4 L bottle at standard pressure and temperature, 6.02×10^{23} molecules exist. This number, 6.02×10^{23}, represents the Avogadro number of molecules in one mole of the gas. To understand this extremely large number, calculate how long it would take to empty the bottle if one million molecules flow out of the bottle each second.

Section 5.2

9. Establish unit ratios for the following desired conversions: (a) seconds to minutes; (b) pounds to slugs; (c) feet to inches; (d) meters to centimeters; (e) amperes to milliamperes (mA).

---

*Challenging problem.

Example: $\frac{\text{seconds}}{1} \times \frac{\text{minutes}}{\text{seconds}} =$ minutes, and the unit ratio is $\frac{1 \text{ min}}{60 \text{ s}}$

10. Perform the following conversions (write unit ratios and show all four steps as outlined in the text):

   a. 7.83 ft to inches

   b. 6′4″ to feet

   c. 5′6″ to inches

   d. 65 statute miles to yards

   e. 110 pounds to slugs

   f. 12 hr to minutes and to seconds

   g. 76 years to seconds

   h. 40 mph to feet per second (see examples 5.5 and 5.6)

   i. $320 \dfrac{\text{mi}}{\text{hr}}$ to $\dfrac{\text{ft}}{\text{min}}$

   $40 \dfrac{\text{mi}}{\text{h}}$ to $\dfrac{\text{ft}}{\text{s}}$

11. In Example 5.2, the result of 4.708 ft when converted to inches does not exactly equal 56.5 in. What is the error in the result due to?

12. How many minutes of study time are contained in a two-year college program? Use 80 weeks at a rate of 25 hours per week.

13. Convert a velocity of 70 mph to feet per minute.

14. Convert the rotational velocity of a gear at 45 revolutions per minute (rpm) to revolutions per second (rps).

*15. The speed of light is 186 000 miles per second and the sun, on average, is 93 million miles from the earth. How long, in minutes, does it take for a photon of light to get from the sun to the earth? Hint: First divide the distance in miles by the speed in miles per second. Cancel the appropriate units and complete your solution with the final conversion.

16. In Example 5.7, the Phantom aircraft carries a total of 19 000 lb of fuel at takeoff. How many gallons of jet fuel does this represent, if the weight density of the fuel is 6.5 lb/gal?

Section 5.3

17. Match the following SI prefixes with the numbers they represent. Refer to Table 5.3 only if a confidence check is necessary. These prefixes and their values must be memorized.

SI Prefix and Symbol		*Numerical Size*
1. kilo-	k	**a.** 0.001
2. nano-	n	**b.** 1×10^{-12}
3. giga-	G	**c.** 1000
4. centi-	c	**d.** 10 000
5. micro-	μ	**e.** 1 000 000
6. milli-	m	**f.** 10^{-9}
7. mega-	M	**g.** 0.01
8. pico-	p	**h.** 10^{-6}
		i. 1×10^{-8}
		j. 1×10^{9}

18. Perform the following conversions (write unit ratios (see inside front cover) and show all four steps as outlined in the text):

   a. 54 m to kilometers

   b. 120 cm to meters

---

*Challenging problem.

    c. 25 mL to liters

    d. 5500 Pa to megapascals.

19. Convert 0.503 meters to centimeters and millimeters.

20. Convert a 37 800 kg tractor-trailer (18-wheeler) to a force in newtons. Express your answer in scientific and engineering notation.

21. The pressure of earth's atmosphere at sea level is 1.013×10^5 newton/m^2. Express this number in kilopascals (kPa).

22. How many millimeters per week does human hair grow? (See Example 5.10.)

23. The proper term for the abbreviation μm is _____ (micron, micrometer).

Section 5.4

24. By using the $1/X$ or X^{-1} (reciprocal) function on the calculator, convert the following unit ratios to a number, other than one, in the numerator.

    a. $\dfrac{1\text{ ft}}{12\text{ in}}$ *Example:* $\dfrac{83.3 \times 10^{-3}\text{ ft}}{1\text{ in.}}$

    b. $\dfrac{1\text{ mi}}{5280\text{ ft}}$ **c.** $\dfrac{1\text{ hr}}{3600\text{ s}}$ **d.** $\dfrac{1\text{ ft}}{0.305\text{ m}}$ **e.** $\dfrac{1\text{ L}}{0.264\text{ gal}}$ **f.** $\dfrac{1\text{ cm}}{1 \times 10^{-2}\text{ m}}$

***25.** If 1 in. is equal to 25.4 mm, what is 1 mm of length equal to in inches?

26. The Soviet SS-N-20 Seahawk missiles, designed for submarine launch, carry eight 500 kiloton MIRVs (*m*ultiple *i*ndependently targetable *r*eentry *v*ehicles). Find the equivalent weight in tons, pounds, and newtons for all eight warheads. (Use the English ton.)

27. If the missile submarine in Problem 26 can dive 60 m in 5 min, find how quickly it dives in feet per minute. How many feet can it dive in 6 min and 34 s?

28. Solve for your mass in slugs and convert slugs to kilograms. Convert your weight in pounds to newtons (see Example 5.19).

29. What is the closest metric wrench size to the 1/2 inch wrench? The answer must be accurate to $+/-$ 1 mm.

30. Convert a speed of 70 mph to km/hr.

31. The tolerance for an automobile part is 0.001 in. to -0.003 in. Convert the tolerance to centimeters, millimeters.

32. Read the micrometers in Figure 5.16. Convert the millimeter reading to inches and the inch reading to millimeters.

33. An auto mechanic uses a torque wrench to torque a bolt to 350 in · lb. What is the torque in N · m? (See Example 5.14.)

***34.** See Figure 5.17. A blood pressure report reads 110 over 76. You find that the reading is in millimeters of mercury (mm Hg). Convert this reading to inches of mercury (in. Hg).

---

*Challenging problem.

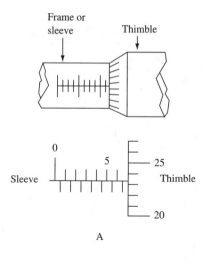

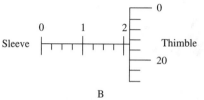

Figure 5.16 Figure for Problem 32. Figure A: 5 = 5 mm. Figure B: 1 = 0.1 in.

Section 5.5

35. Convert the following areas and volumes:

    a. 60 cm² to m²

    b. 40 ft² to in.²

    c. 40 000 ft² to mi²

    d. 28 gal to liters

    e. 4.66 L to cubic centimeters

    f. 0.000 078 L to milliliters.

36. A box is 2 ft × 1 ft × 4 ft. How many cubic inches does it contain? How many cubic centimeters does it contain?

*37. A pint is what percent (%) of a U.S. gallon? A liter?

38. How many quarts are contained in a bottle at the grocery that contains 946 mL?

39. A container at the grocery reads 1 pt. 6 oz. How many mL of product does it contain? Hint: There are 16 ounces to a pint and 8 pints to a gallon.

40. Convert 28 mi/gal to mi/L.

---

*Challenging problem.

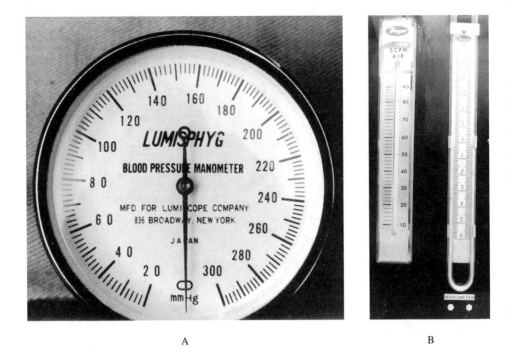

Figure 5.17 Both types of mercury (Hg) manometers can measure blood pressure. One reads in millimeters of Hg and the other in inches of Hg.

*41. While traveling in Britain you find that gasoline or petrol is 50 cents (U.S.) per liter. Convert this cost to dollars per gallon.

42. If power $(P) = \dfrac{F \times s}{t}$ and F is in units of $\dfrac{kg \cdot m}{s^2}$, s is in units of meters, and t is in units of seconds, what are the units P is expressed in?

Section 5.6

43. Complete the following steel wire gage table by filling in the blank spaces.

Gage	*Decimal Equivalent (in.)*	*Metric Equivalent (mm)*
2/0	0.331	_____
1/0	0.306	_____
1	_____	7.19
2	_____	6.65

44. How tall is a 5′11″ person in millimeters? Round off the accuracy of your answer to the relative accuracy of the measurement in inches. Consider the accuracy of the inch measurement to be ±1/2 in., or ± _____ mm.

---

*Challenging problem.

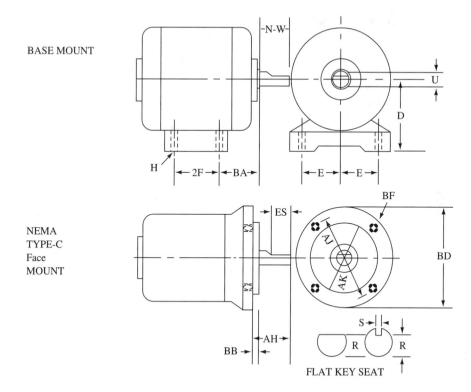

Figure 5.18 Figure for Problem 45.

45. See Figure 5.18. Convert the following National Electrical Manufacturer's Association (NEMA) dimensions to metric equivalents. All dimensions are in inches; NEMA type-C face mount.

NEMA Frame	AH	AJ	AK	BB Min.	BD Max.
42	1.31 ____	3.75 ____	3.0 ____	0.16 ____	5.00 ____
48	1.69 ____	3.75 ____	3.0 ____	0.16 ____	5.62 ____
56	2.06 ____	5.88 ____	4.5 ____	0.16 ____	6.50 ____

For Problems 46–50 *perform the measurement* in inches or feet and inches, and then convert to the appropriate metric unit. Show all conversions. Record the uncertainty of the final answer (Section 5.6).

46. A dollar bill's length and width.

47. The diameter of a dime.

48. An 8½ × 11 sheet of paper's length and width.

49. The length and width of a standard door.

50. The area of the cover of this textbook in square millimeters. Convert the square millimeters to square inches.

Selected Readings

American National Standards Institute. *ISO Standard 1000.* 1987.

American Society for Testing and Materials. *Standard for Metric Practice E 380.* March 1, 1984.

Military Handbook 52 A, 1984.

National Bureau of Standards.* *Special Publication 304A & 330.* 1986.

National Bureau of Standards.* *Weights and Measures Standards of the United States: A Brief History.*

---

*The National Bureau of Standards for the United States was renamed the National Institute of Standards and Technology (NIST) in 1988.

6

Right-Triangle Trigonometry and Geometry for Technologists

The right triangle has been one of history's most used engineering tools. In 2000 B.C., Egyptians were using the magic "3-4-5" triangle to build pyramids. Over 2500 years ago, the Greek philosopher Pythagoras formulated the Pythagorean theorem, which gave further definition to the useful right triangle.

Today, technologists use the right triangle to calculate such diverse relationships as vector forces on beams and walls, electrical currents and voltages that are out of phase, and the distance a machine tool will travel when given instructions to move along X and Y coordinates.

You will find the right triangle to be indispensable during your career; you will also find it easy to understand. This section will give you an overview of right-triangle relationships. Later, you will encounter mathematics courses that will lend further understanding of this important mathematical tool.

6.1 Right-Triangle Relationships

Carpenters still use the magic 3-4-5 triangle (Figure 6.1) to make "square" corners (right-angled corners). The Pythagorean theorem can be used to prove that a triangle with sides equaling $3 \times 4 \times 5$ will form a right angle opposite the side of length 5.

EXAMPLE 6.1

The **Pythagorean theorem** is written as $c^2 = a^2 + b^2$. Prove that a $3 \times 4 \times 5$ triangle will form a right angle (a 90° angle).

Solution First, sketch a right triangle (one angle is equal to 90°) with the 90° angle to the right of the drawing. The longest side across from the right angle, the hypotenuse, is labeled c. The vertical side will be labeled a and the horizontal side b. Finally, dimension the drawing as shown in Figure 6.2.

153

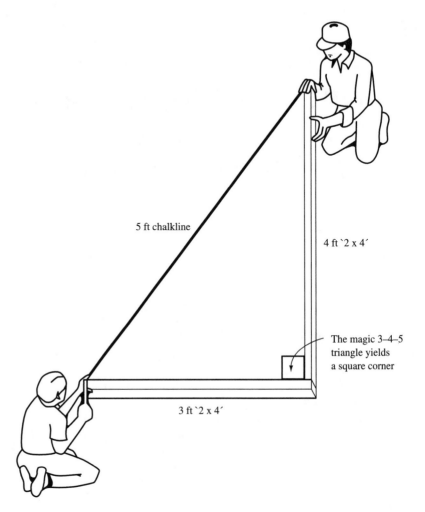

5 ft chalkline

4 ft `2 x 4´

The magic 3–4–5 triangle yields a square corner

3 ft `2 x 4´

Figure 6.1 Carpenters laying out a square corner.

Next, use the Pythagorean theorem to show that

$$c^2 = a^2 + b^2 = 4^2 + 3^2 = 25$$

So, $c^2 = 25$, and

$$c = \sqrt{25} = 5$$

The "magic triangle" is proven to be a right triangle by the Pythagorean formula.

EXAMPLE 6.2 A mechanical engineering technician programs a machine tool to move 0.004 in. in the $+X$ direction, and 0.007 in. in the $+Y$ direction. How far does the tool move?

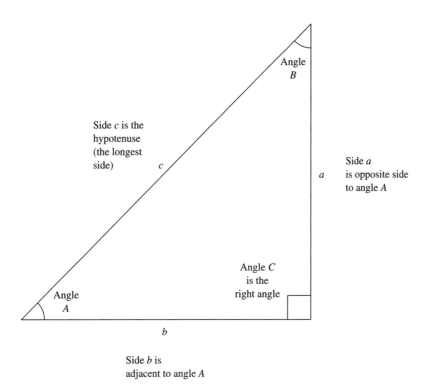

Figure 6.2 The right triangle, in standard position, has sides and angles labeled in a conventional format.

Solution   First, the sketch is drawn in standard form (Figure 6.3).

Note: The sketch is labeled in a standardized format followed in many texts. The angles are identified by capital letters and are opposite the sides carrying the same (but not capitalized) letters. Think of side *a* as the *a*ltitude and side *b* as the *b*ase of the standard triangle.

The calculation is

$$c^2 = (0.007)^2 + (0.004)^2 = 6.5 \times 10^{-5}$$

$$c = \sqrt{6.5 \times 10^{-5}} = 8.06 \times 10^{-3} = 0.00806 \text{ in. or } 8.06 \text{ mil in.}$$

Another relationship that will aid in solving right triangles is the complementary-angle relationship. For any triangle—not just right triangles—the three angles will always total 180°. Because one angle of a right triangle is 90°, the other two angles must total 90°. Two angles that total 90° are **complementary.** Therefore, the two angles of a right triangle, other than the right angle, are complementary, or will always add to 90°.

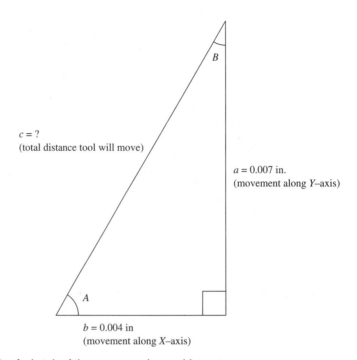

$c = ?$
(total distance tool will move)

$a = 0.007$ in.
(movement along Y–axis)

A

$b = 0.004$ in
(movement along X–axis)

Figure 6.3   A sketch of the programming problem.

EXAMPLE
6.3

In Example 6.2, angle A is 60.3°. Find angle B.

Solution   By the complementary relationship, angle $B = 90° -$ angle A. Therefore,

$$B = 90° - 60.3° = 29.7°$$

The two relationships discussed—the Pythagorean theorem and the complementary-angle relationship—will, when combined with the **trigonometric functions,** allow you to solve any right triangle when given two measurements on the triangle. The trigonometric functions for the standard triangle (Figure 6.4) are the sine, cosine, and tangent. On the calculator they are abbreviated as sin, cos, and tan. The trigonometric functions for angle A are defined as

$$\sin A = \frac{\text{opposite side}}{\text{hypotenuse}} = \frac{a}{c},$$

$$\cos A = \frac{\text{adjacent side}}{\text{hypotenuse}} = \frac{b}{c},$$

$$\tan A = \frac{\text{opposite side}}{\text{adjacent side}} = \frac{a}{b}$$

These ratios are written for angle A. See Problem 4 at the end of this chapter and write the ratios for angle B. You may wish to refer to Appendix C for additional help with trigonometric relationships.

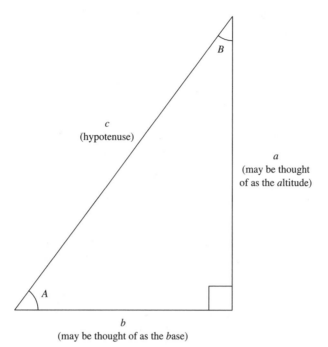

Figure 6.4 The trigonometric functions are defined for the right triangle.

Because the numerator and denominator of each ratio represent the same quantity, i.e., distance, the units cancel. The trigonometric functions, therefore, are unitless ratios.

These ratios are the same for all similar triangles. A similar triangle is one having similar, or equal, angles. In Figure 6.5 three similar triangles are superimposed. The drawing is to scale. Using a ruler, confirm that the ratio of the sides—a′/b′, a″/b″, a′ ″/b′ ″—of the similar triangles are all equal to one.

Because the ratios of the sides in the figure are the same, a table can be developed so you can look up the values of these functions. You can also easily solve the values of the trigonometric functions with your calculator.

EXAMPLE 6.4

Use your calculator to solve for the sine (sin), cosine (cos), and tangent (tan) functions of the angles 0°, 30°, 45°, 60°, and 90°. Construct a table (Table 6.1) to see how the trig functions increase or decrease in value as the angle of interest becomes larger.

Note: *Before attempting this example, confirm that your calculator is in the degree mode, not in the radian or grad mode.*

For single line calculators, press the ⌈DRG⌉
ters ⌈Deg⌉ appear on the screen. For the TI-85 calculator press the ⌈2nd⌉ ⌈Mode⌉ key and select ⌈Deg⌉ by using the arrow keys at the top right of the keyboard. (The radian and grad modes represent other angular measurement units. *Your instructor may wish to*

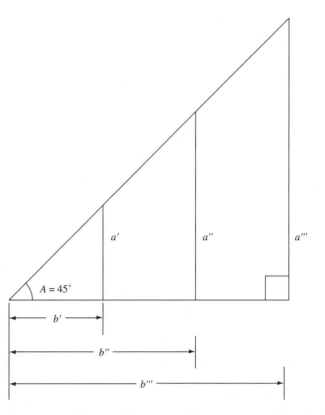

Figure 6.5 Similar triangles have similar angles and the sin, cos, and tan ratios are equal. For instance, $\tan A = a'/b' = a''/b'' = a'''/b''' = 1$ (for angle A of 45°, as shown in the figure).

introduce the radian measure at this time, for it is used frequently in engineering calculations. *This text, however, will always use the degree mode.* The more familiar degree units will speed your understanding of the fundamental principles of the trigonometry.) Press [ENTER] to fix the calculator in the degree mode, and use [EXIT] to return to the home screen.

Solution Use the following sequence of steps, on your calculator, to solve for the sine, cosine, or tangent of each angle.

enter angle—push [sin] [cos] or [tan] keys—read display

(TI-85 users will enter the [sin] [cos] [tan] keys first and then enter the angle.)

As mentioned in Chapter 4, this text is not meant to replace your calculator manual. Please refer to it now and try some of the examples in the manual to familiarize yourself with the function keys.

What can we observe from the table we constructed? Here are a few observations.

Table 6.1 Partial Table of the Values for Trig Functions*

Calculator Key Presses	Angle A	Sin A	Cos A	Tan A
0 [sin] = 0.000				
0 [cos] = 1.000	0°	0.000	1.000	0.000
0 [tan] = 0.000				
30 [sin] = 0.500				
30 [cos] = 0.866	30°	0.500	0.866	0.577
30 [tan] = 0.577				
45 [sin] = 0.707				
45 [cos] = 0.707	45°	0.707	0.707	1.000
45 [tan] = 1.000				
60 [sin] = 0.866				
60 [cos] = 0.500	60°	0.866	0.500	1.732
60 [tan] = 1.732				
90 [sin] = 1.000				
90 [cos] = 0.000	90°	1.000	0.000	Error!
90 [tan] = Error!				

* This text does not provide a complete table of trigonometric functions. Your scientific calculator has the functions stored in memory.

1. As angle A increases, sin A *increases* from 0.000 to 1.000.
2. As angle A increases, cos A *decreases* from 1.000 to 0.000.
3. The sine and cosine of 45° are equal. Can you explain why? (see Problem 5 at the end of this chapter.)
4. As angle A increases, tan A increases from 0 to 1 at 45° and to an error at 90°. Can you explain why an error was displayed at 90°? (see Chapter 4, Example 4.16, and Problem 6 in this chapter.)
5. The sine and cosine of complementary angles (see 30° and 60° in Table 6.1) are equal.

With the calculator you are also able to find an angle for a particular function when the value of the ratio is given. You can use Table 6.1 to confirm how your calculator does this **inverse trigonometric function.**

Note: Do not confuse the inverse functions on the calculator with the reciprocal functions, known as the secant, cosecant, and cotangent.

For instance, the inverse sine (also read as the arc sine) function for 0.500 is found in Table 6.1 to be a 30° angle. What is the inverse cosine function (arc cosine) for 0.500? On the calculator the inverse functions are accessed by pressing [INV] and the function or by pressing [2nd F] [cos⁻¹] (the superscript −1 does not mean the same thing as when it is used as a power of a number; here it means inverse cosine, or arc cos). Confirm the inverse functions for all the other ratios in Table 6.1 by using your calculator. If you cannot, consult your operator's manual or see your instructor.

EXAMPLE
6.5

Find the inverse tangent function for 1.750.

Solution We find the inverse tangent of 1.750 ($\tan^{-1}$ 1.750) by pressing

1.750 [tan⁻¹] = 60.3° (some calculators use 1.750 [INV] [tan] [=])

TI-85 users will enter the [sin⁻¹] [cos⁻¹] [tan⁻¹] keys before entering the ratio first, for example,

[2nd] [tan⁻¹] 1.75 [ENTER]

Notice that the arc tangent, or inverse tangent, for 1.750 must result in an angle greater than 45°. Above 45°, the tangent function takes on a value greater than one. See Table 6.1 in Example 6.4 to confirm this.

The relationships for the right triangle in this section include

- Pythagorean theorem, or $c^2 = a^2 + b^2$,
- complementary angles, or angle A + angle $B = 90°$ in a right triangle,
- the trigonometric functions: sine, cosine, and tangent, and
- the inverse trigonometric functions: [sin⁻¹], [cos⁻¹], [tan⁻¹].

The next section will provide applications. The applications will allow you to practice these relationships, become more familiar with them, and realize how important right-triangle trigonometry could be in your career. Refer to **Appendix C** as you solve the example problems that follow. You should be able to solve for all of the missing components of each of the right triangles in Appendix C.

6.2 Right-Triangle Applications

The following exercises will introduce the type of right-triangle problems that you will encounter in the engineering technologies. The examples all involve triangles placed in standard position. At first, the triangles will be used only to solve for distance; later they are used to solve for other quantities such as force, velocity, and electrical potential. If you have trouble deriving algebraic equations, use Figure 6.6 to construct the equations necessary to solve for the unknown quantity of interest. Merely cover the unknown with your finger to see the relationship you must solve for. The Greek letter θ (theta) represents any angle—i.e., angle A or B.

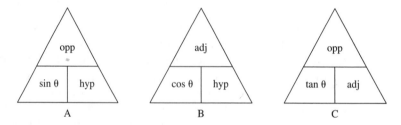

Figure 6.6 Using the three partitioned triangles, cover the unknown to be solved for, and the relationship of the known quantities can be seen.

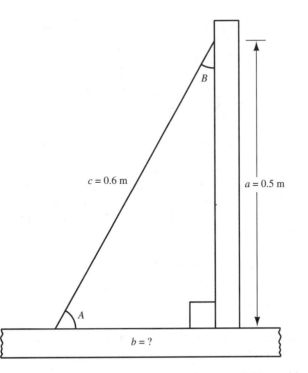

Figure 6.7 A brace is to be manufactured to support a perpendicular weld.

EXAMPLE
6.6

A triangular metal brace must be manufactured to support two welded pieces perpendicular to each other. If the longest side (hypotenuse) is 0.6 m and the support
side is 0.5 m, solve for the two angles and the length of the base (Figure 6.7).

Solution   First, check to be sure your calculator is in the degree mode. Solve for angle
A by using the sine function:

$$\sin A = \frac{\text{opp}}{\text{hyp}} = \frac{0.5}{0.6} = 0.833$$

and

$$\sin^{-1} 0.833 = \text{angle } A = 56.4°$$

Using the complementary relationship,

$$\text{angle } B = 90.0° - 56.4° = 33.6°$$

Using the Pythagorean theorem,

$$b^2 = c^2 - a^2$$

and

$$b = \sqrt{c^2 - a^2} = \sqrt{0.36 - 0.25} = 0.332 \text{ m}$$

In trigonometry, checkbacks are plentiful, and in this case you could use

$$\tan A = \frac{opp}{adj} = \frac{0.5}{b}$$

and

$$b = \frac{opp}{\tan A} = \frac{0.5}{\tan 56.4°} = \frac{0.5}{1.505} = 0.332 \text{ m}$$

The checkback assures us that previous calculations were accurate.

EXAMPLE 6.7 A surveyor checks the grade (slope) of a mountain highway. If the average grade is 8°, what is the length of highway for every mile of horizontal distance? What is the rise of the highway for each horizontal mile of distance? What is the percent of the grade?

$$\% \text{ grade } = \frac{rise}{run} \times 100$$

See Figure 6.8.

Solution Because the adjacent side and angle A are known, the cosine function is selected:

$$\cos A = \frac{adj}{hyp}$$

and (use the middle triangle of Figure 6.6)

$$c \text{ (hyp)} = \frac{adj}{\cos A} = \frac{1}{\cos 8°} = \frac{1}{0.990} = 1.01 \text{ mi}$$

Therefore, for every mile of horizontal distance covered, 1.01 mi must be asphalted. Use the Pythagorean theorem to find the rise, or side *a*.

$$a^2 = c^2 - b^2$$

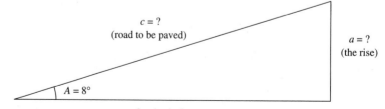

$c = ?$
(road to be paved)

$a = ?$
(the rise)

$A = 8°$

$b = 1$ mi of run

Figure 6.8 The average grade of the mountain road is 8°.

and

$$a = \sqrt{1.01^2 - 1^2} = 0.142 \text{ mi}$$

The road rises 0.142 mi for every mile of run (horizontal distance). The percent grade equals

$$\% \text{ grade} = \frac{\text{rise}}{\text{run}} \times 100 = \frac{0.142}{1} = 14.2$$

Is there a trigonometric function that could solve for the percent grade more directly?

EXAMPLE 6.8

A ceiling lamp (8 ft ceiling) must illuminate a 1.6 m surface on a table 0.70 m high. At what angle will you design the shade to assure proper illumination? Add 1 ft of height for the virtual light source.

Solution   Note that one-half of the triangle in Figure 6.9 is a right triangle with a side *a* of 8 ft + 1 ft = 9 ft

Converting the feet to meters the altitude (*a*) is

$$9 \text{ ft} \times \frac{0.304 \text{ m}}{1 \text{ ft}} = 2.74 \text{ m} - 0.7 \text{ m (the height of the table)} = 2.04 \text{ m}$$

Side *b* or the base is one-half of the diameter of the table (0.8 m). The angle of the shade will correspond to angle *B*. Therefore

$$\tan B = \frac{\text{opp}}{\text{adj}} = \frac{b}{a} = \frac{0.8}{2.04} = 0.392$$

and

$$\tan^{-1} 0.392 = 21.4°$$

Of course there will be a fringing effect, but the lampshade designed to be at an angle of 20° from the perpendicular will provide a minimum diameter of the desired illumination.

The mechanical engineering technologist must be able to set up machine tools to cut materials at specified angles. For instance, a lathe tool may have to be adjusted to cut a cone-shaped piece of stock. The angle of the compound rest on the lathe is then adjusted to that angle before turning begins (Figure 6.10A).

EXAMPLE 6.9

Calculate the angle to set the compound rest for the lathe so the part can be manufactured (angle *A*).

Solution   Use the left-hand side of the cross section to calculate angle *A*. Side *a*, or the opposite side (the height of the cone), and side *b*, or the adjacent side (use only one-half of the base), are given. Therefore, the tangent function will be used to solve for angle *A*:

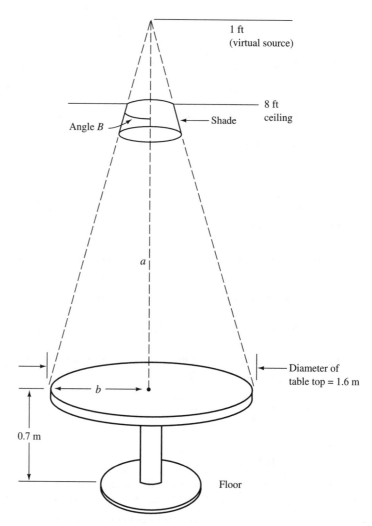

Figure 6.9 The ceiling lampshade is designed to illuminate a table of 1.6 m diameter.

$$\tan A = \frac{\text{opp}}{\text{adj}} = \frac{0.495}{\frac{0.560}{2}} = \frac{0.495}{1} \times \frac{2}{0.560} = 1.77$$

and

$$\tan^{-1} 1.77 = 60.5°$$

Note: The compound rest must be set at the complementary angle, or the angle of the top of the cone to its longitudinal axis (29.5°). Refer to Figure 6.10B.

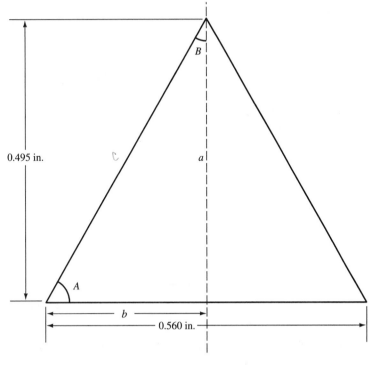

0.495 in.

B

a

A

b

0.560 in.

A

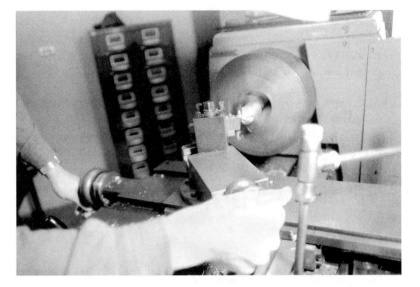

B

Figure 6.10 (A) The cone-shaped object (cross-sectional view) must be turned on the lathe. (B) Photo of compound rest for lathe.

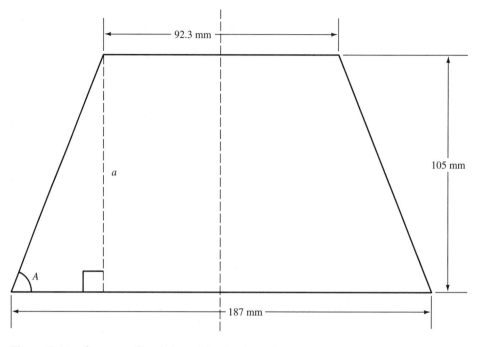

Figure 6.11   Cross-sectional view of the frustum of a cone.

EXAMPLE
6.10

Figure 6.11 depicts a portion of a cone whose upper part has been cut off by a plane parallel to the base (i.e., frustum of a cone). Solve for the angular set of the lathe's compound rest.

Solution   Again, work only with angle *A*. Because the base of the frustum is parallel to the base of the cone, a perpendicular to the base can be drawn to intersect the top at the left-hand edge. This forms a right triangle with side *a* equal to the height of the frustum above the base and the base of the triangle (side *b*) equal to one-half the base minus one-half the top ($187/2 - 92.3/2 = 47.4$).

$$\tan A = \frac{\text{opp}}{\text{adj}} = \frac{105}{47.4} = 2.22$$

and

$$\tan^{-1} 2.22 = 65.8°$$

From the lathe bed, the compound rest will be adjusted to the complementary angle of $24.2°$.

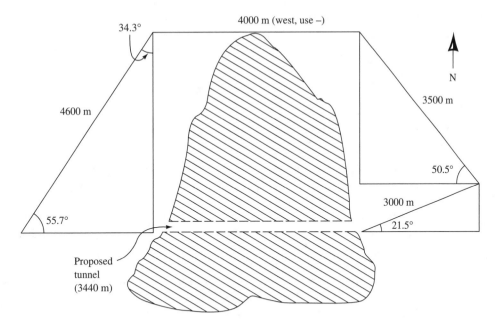

Figure 6.12 Cross section of mountain for proposed tunnel.

6.3 Vector Applications

A surveyor can use right-triangle trigonometry to locate the openings of a tunnel on both sides of a mountain (Figure 6.12). Construction of the tunnel can then begin from both sides at the same time, halving the time required for boring and shoring the tunnel.

EXAMPLE 6.11

A tunnel is to be constructed on an east-west line through a mountain. Find the exact point, on both sides of the mountain, where the work must begin.

Solution The surveyors begin on the east side and triangulate as shown in Figure 6.12. We will use the following transformations of the trigonometric functions and Table 6.2 to keep track of the movements in the north and east directions (use minus signs for south and west directions). The equations with the table illustrate **method of components,** a powerful technique used to analyze complex vector quantities in physics and many technical courses, e.g., strength of materials and AC circuits.

movement east = hyp × cos θ

movement west = hyp × sin θ

See Figure 6.6.

Table 6.2 Constructing the Tunnel

Distance	Direction	Movement East (m)		Movement North (m)	
3000	21.5° NE	3000 cos 21.5° =	2790	3000 sin 21.5° =	1100
3500	50.5° NW	=	−2230	=	2700
4000	west!	=	−4000	=	0
4600	55.7° NE	=	−2590	=	−3800
2590	east!	=	2590	=	0
		Total	−3440 (length of tunnel)		0

In Table 6.2, movements are tabulated in meters north and east. All calculations are rounded to three-significant-figure accuracy. The missing equations in the two columns are to be completed by the student (see Problem 20). The total is the total displacement from where the surveyor began. The −3440 m represents a west direction and is the length of the proposed tunnel. The zero movement north confirms that the tunnel will be built on an east-to-west plane.

The information given in Figure 6.12 is known as polar coordinate information. A polar coordinate identifies a vector by using an angle and a distance. Polar coordinates can be symbolized by $\boxed{R/\theta}$, where R is the length of the vector (a radius) and $\angle\theta$ is the angle from the $+X$ axis. For instance, the 3500 m vector could be written in the form $3500\angle129.5°$ (use $180° - 50.5°$ to find the angle from the $+X$ axis).

The scientific calculator will automatically convert from polar ($R\angle\theta$) to rectangular (X,Y). The calculator keys used are $\boxed{R/\theta}$ and $\boxed{X,Y}$ (or $\boxed{P\text{-}R}$). For the TI-85, the CPLX (Complex Number) Menu must be accessed. The key presses for the fourth vector in Table 6.2 are (calculator must be in degree mode)

$$\boxed{(} \ 4600 \ \boxed{2nd} \ \boxed{\angle} \ 55.7° \boxed{)} \ \boxed{2nd} \ \boxed{CPLX} \ \boxed{MORE} \ \boxed{\blacktriangleright REC}$$

The result is read as (2.592E3, 3.800E3). Converting from Engineering Notation to Floating Point, rounding to three significant figures, and considering the actual displacement, the result is −2590 (movement West instead of East) and −3800 (movement South instead of North). From Figure 6.12 you can see that the displacement of the vector is actually west and south, which results in the minus signs. More powerful methods of determining vector direction will be discussed in your trigonometry course.

With the automatic conversion on the calculator, you may confirm the results obtained in Table 6.2 (have you already completed the table?). You should always consult your owner's manual for the exact conversion procedure.

The displacements in Example 6.11 are vector quantities. A **vector quantity** is defined as a quantity possessing both magnitude and direction. Forces may be represented by vectors because all forces possess both magnitude and direction. When two vectors act at the same point, they are concurrent. See Figure 6.13.

EXAMPLE 6.12 In Figure 6.13, solve for the total downward force on the beam, represented by the two force vectors. Pounds are used for the units, but newtons in the international system could be used as well.

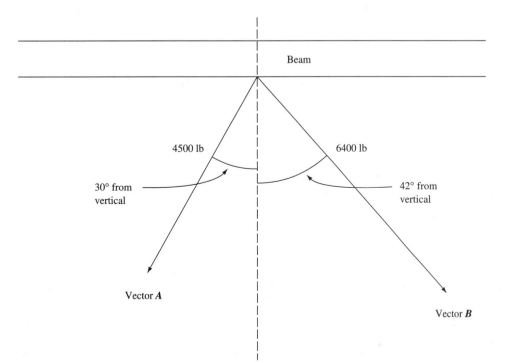

Figure 6.13 Two concurrent force vectors act on a beam. The two vectors could represent loads supported by the beam, such as ceiling fixtures and overhead cranes.

Solution   We will solve only for the vertical components of the vectors. Once these *Y* components are determined, we may add them together to find the total downward loading of the beam.

Because the angles are defined from the vertical, the *Y*, or vertical component, of the vectors may be solved for by using the cosine function (the adjacent side to the given angles is the vertical, downward component):

$$\cos \theta = \frac{\text{adj}}{\text{hyp}}$$

and, after transposing

$$\text{hyp} \times \cos \theta = Y$$

$$\text{vector } A = 4500 \times \cos 30° = 3900 \text{ lb, vertically down}$$

$$\text{vector } B = 6400 \times \cos 42° = 4760 \text{ lb, vertically down}$$

$$\text{Total} = 8660 \text{ lb, vertically down}$$

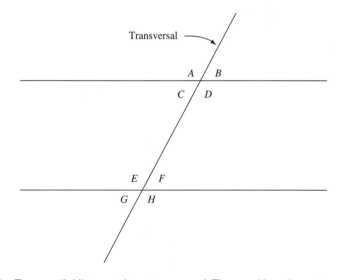

Figure 6.14 Two parallel lines cut by a transversal. The resulting alternate angles, alternate interior angles, and alternate exterior angles are equal.

You will appreciate how important it is to define loads on systems we all depend on for our safety. By using vector addition and method of components, you can easily determine many loads.

In Example 6.12 we solved for the downward component of the two vector forces. If we had wished to solve for the resultant total force, we could have used the vector-triangle method. The vector-triangle method uses a graphical solution for adding vectors. The **vector-triangle method** is "to add two vectors, move the tail of one of the vectors to the head of the other—the **resultant (R)** is the vector drawn from the head of the first vector to the tail of the last."

To begin our discussion, consider the angles in Figure 6.14. The angles result from cutting two parallel lines by a transversal. It is fairly easy to see that angles A and D, B and C, E and H, and F and G, all known as **alternate angles,** are equal. The parallel lines yield even more equalities. The alternate interior angles and the alternate exterior angles are equal. By inspection you should be able to determine these angles; they are

> alternate angles = A and D, B and C, E and H, F and G
>
> alternate interior angles = C and F, D and E
>
> alternate exterior angles = A and H, B and G

This geometric concept aids us in visualizing the movement of vectors. You may move a vector to any location if you maintain the magnitude (length) and keep it parallel to its original position.

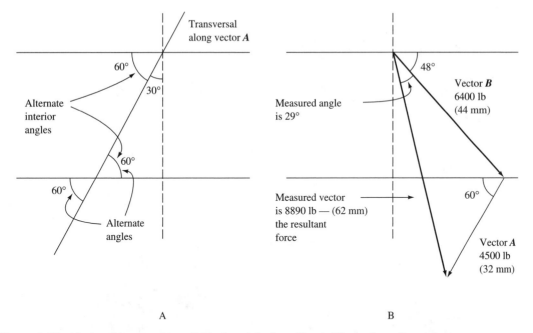

Figure 6.15 Vector **A** is moved parallel to its original position in Figure A, so it may be added to vector **B** (Figure B). The resultant can then be determined by measurement (for this drawing, use 0.7 mm = 100 lb force).

EXAMPLE
6.13

Graphically add the two vectors in Figure 6.13 by using the vector-triangle method.

Solution We sketch the two vectors before (Figure 6.15A) and after (Figure 6.15B) vector *A* has been moved to the head of vector *B.* Parallel lines are drawn to aid in maintaining vector *A* parallel to its original position. Vector *A* is first extended as a transversal to the parallel lines and later moved to the head of vector *B,* maintaining the angle of the transversal. The resultant vector is drawn to scale. You may solve for the resultant by using protractor and ruler to obtain 8890 lb at 77° from the horizontal.

The electronics technician and technologist must be familiar with **alternating current (AC)** and **voltage,** such as the current and voltage supplied to our homes (i.e., 120 V, 60 cycles/s). AC, as contrasted with DC, or direct current, takes the form of a sine wave as it continually changes in magnitude. This is due to the action of the **generator.** A simple model of a generator is seen in Figure 6.16A. The voltage induced in the coil of the conductor rotating in the magnetic field is proportional to the rate of change of the lines of force cutting the conductor (determined by Faraday in 1831). From the simple model, you can visualize how the number of lines of force cut by the conductor per second is equivalent to the magnitude of the current or voltage from

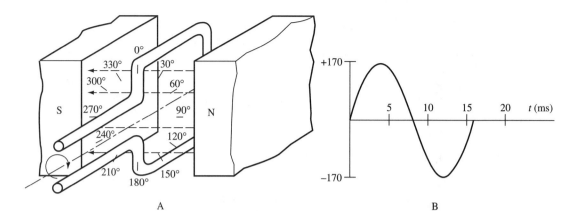

A   B

C

Figure 6.16 Household voltage changes constantly with time as the generator's conductor sweeps through the magnetic lines of force at differing rates.

the generator. For example, when the conductor is oriented as shown, its motion would cut across no lines of force, yielding an output of 0, as shown in Figure 6.16B. At 90°, it would cut across a maximum number of force lines per second, yielding a maximum output as shown in Figure 6.16B.

The peak voltage of the 120 V service is actually 170 V (120 V is the effective DC component of the sine wave), and the sine wave completes one cycle in 0.0167 s. Simply invert 60 cycles per second (the frequency, f) to solve for the period (T) of the wave in seconds per cycle:

$$T = \frac{1}{f} = \frac{1}{60 \text{ cycles/s}} = 0.0167 \text{ s/cycle}$$

The **sine wave** is drawn for one full cycle in Figure 6.16B. The technologist is measuring the sine wave on the oscilloscope in Figure 6.16C.

EXAMPLE
6.14

Solve for the instantaneous voltage (e) of the sine wave in Figure 6.16 at $t = 3$ and 7 ms.

Solution First we must solve for the angles at the times of interest. Note how the use of unit conversion addressed in Chapter 5 aids in this calculation. All that remains after the angle is determined is to find the sine of the functions and multiply by the peak voltage. The function (equation) is described as $e = V \sin \theta$. At $t = 3 \times 10^{-3}$ s,

$$\theta = \frac{3 \times 10^{-3} \text{ s}}{1} \times \frac{1 \text{ cycle}}{16.7 \times 10^{-3} \text{ s}} \times \frac{360 \text{ deg}}{1 \text{ cycle}} = 64.7°$$

At 3 ms, we have rotated through 64.7°. The instantaneous voltage is $e = 170 \sin 64.7°$ = 154 V. You can see from Figure 6.16B that at 3 ms the curve would be near V_{max}.

At $t = 7 \times 10^{-3}$ s, $\theta = 151°$. Use the following keystrokes to solve for the sine of 151°:

151 ⎡sin⎤⎡=⎤ 0.486

and

$e = 170$ ⎡x⎤ 0.486 ⎡=⎤ 82.7 V

The calculator will automatically solve for the function of any angle up to approximately $\pm 10^{10}$ degrees, a very large angle. It is important to note that the sine of the angle takes on a negative value for some angles (e.g., 181° to 359°). As an example, at 13 ms, $\theta = 280°$, and the sine of the angle is -0.984. What would be the instantaneous value of e in this case?

6.4 Geometry Applications for the Technologist

Shapes other than the useful right triangle are encountered in the technologies. Many of these shapes are two-dimensional—they can be drawn on a flat plane. Most of these shapes can be classified as

1. circles,
2. triangles, including all triangles and not just the right triangle,
3. quadrilaterals, such as squares, rectangles, parallelograms, and trapezoids (see Figure 6.18), or
4. other polygons (e.g., figures having more than four sides or angles).

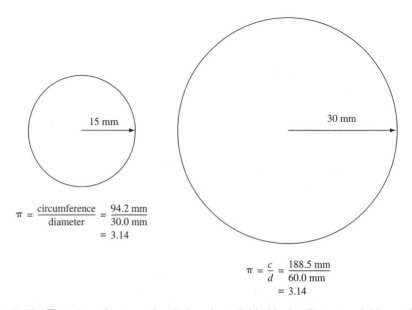

$$\pi = \frac{\text{circumference}}{\text{diameter}} = \frac{94.2 \text{ mm}}{30.0 \text{ mm}}$$
$$= 3.14$$

$$\pi = \frac{c}{d} = \frac{188.5 \text{ mm}}{60.0 \text{ mm}}$$
$$= 3.14$$

Figure 6.17 The circumference of a circle, when divided by its diameter, yields a unitless ratio of 3.14. The Greek letter π (pi) represents this ratio, good for all circles.

All two-dimensional figures are discussed under the general heading of **plane geometry.**

In a formal geometry course, theorems and proofs are offered to solve for the angles, areas, and perimeters of these figures. This text will offer some of the resultant equations and understandings without formal proofs. The proofs that will be offered are designed to allow you to conceptualize how such measures as area and perimeter are determined.

Circles

First, consider the geometry of the circle. The Greek letter π (pi, pronounced like pie) is central to the understanding of the geometry of the circle. Refer to Figure 6.17, where two circles of different size are drawn. Both have a circumference (length around the edge of the circle) 3.14 times greater than the diameter. The 3.14 is the constant π. On a ten-digit calculator, pressing the $\boxed{\pi}$ key will display the number 3.141 592 654.

Pi has an infinite number of digits. Computer programmers compete to see who can obtain the greatest number of digits. Taking pi to an incredible number of decimal places is also a way to troubleshoot supercomputers, exposing errors in a supercomputer's hardware, software, or memory. The current record for pi, set by a Cray-2 (see Figure 8.2), is a stream of digits 600 miles long!

The constant π was discovered by the ancients, but you can discover it for yourself. To accomplish this rediscovery, use a drinking glass and a string. Use a string length equal to the circumference of the rim of the glass. Carefully measure the length of the string and divide this value by the diameter of the rim. The ratio will not be as accurate as that obtained on a ten-digit calculator, but it will approximate the value of 3.1. This measurement exercise will also acquaint you with the difficulties of accurate measurement.

EXAMPLE 6.15

A bakery must cover the top of 8 in. cakes with a special topping. What is the circumference (c) and area (A) of the top of each cake to be covered?

Solution   The circumference equation can be determined by rearranging the literal equations in Figure 6.17.

$$c = \pi d \text{ (or } 2\pi r)$$

Therefore

$$c = \pi \times 8 \text{ in.} = 25.1 \text{ in. } (\pi \text{ is a unitless ratio})$$

Circular area (A) is also calculated by using the constant π. The equation is

$$A = \pi r^2$$

Substituting our known value (the 8 in. diameter is equal to a 4 in. radius), we find

$$A = \pi(4 \text{ in.})^2 = 16\pi \text{ in.}^2 = 50.3 \text{ in.}^2$$

Determining the area to be covered will aid the bakery in determining how much topping to order for weekly production runs of the cakes. It will be able to maintain an adequate inventory of topping.

Quadrilaterals

A *quadrilateral* is a four-sided figure. Some of the more commonly encountered quadrilaterals are shown in Figure 6.18. The algebraic expressions in the figure are used to solve for the area of each type of quadrilateral. Note that a *trapezoid* has *exactly* one pair of parallel sides and the height (h) is always the distance between the two parallel sides. See Problem 36 for experience with quadrilateral applications.

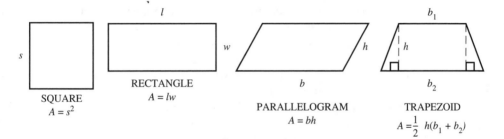

Figure 6.18   A few of the more familiar quadrilaterals found in business and industry.

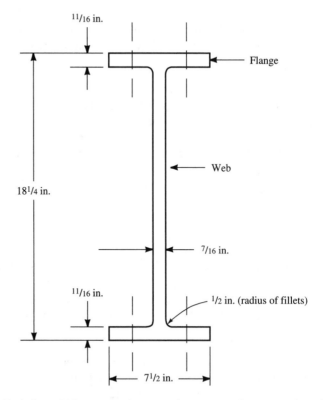

Figure 6.19 End view of I-beam used to manufacture part for power shovel. To calculate the weight of one of these parts, first calculate the area of the end view.

We will use a typical industrial example of a scrap or rework calculation to illustrate the use of geometry in manufacturing. *Scrap* records allow a company to keep track of how much expense is being incurred as waste or to calculate the worth of the scrap generated when it is sold to the scrap yard. *Rework* records show how much material is being reworked or used up—e.g., if I-beams are being used to manufacture a part, the rework calculation shows how many I-beams must be kept in inventory. Rework calculations, then, are used to maintain the manufacturer's *material stores inventory* at adequate levels.

In Figure 6.19, the U.S. customary system is used to dimension the end view of a part to be manufactured. The part is used by a heavy machinery manufacturer that builds power shovels. The part is cut from a standard structural I-beam. The designation for the standard I-beam is $W18 \times 60$. This designation means the nominal depth (see end view) is approximately 18 in. (18.25 in the example) and the nominal weight per foot is 60 lb. You may find the orthographic drawing of the complete part in Figure 6.25 complicated at first, but after working through the examples that follow, every aspect of the drawing will become familiar to you.

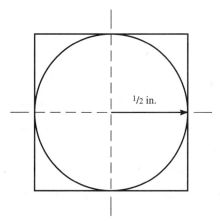

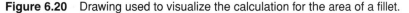

$^1/_2$ in.

Figure 6.20 Drawing used to visualize the calculation for the area of a fillet.

EXAMPLE
6.16

Using the end view of the part in Figure 6.19, calculate the area of the flange and web (we will consider the area of the fillets in the following example).

Solution Both flange and web are rectangles, and length times width is the only calculation required (see Figure 6.18). All common fractions will be rounded to one-thousandth of an inch and all answers will be expressed in decimal fractions accurate to one-hundredth of a square inch.

Area of each flange $= (7.500 \times 11/16) = (7.500 \times 0.688)$

Area of both flanges $= 2(7.500 \times 0.688) = 10.32$ in.^{2}

Area of web $= [18.250 - 2(11/16)] \times 7/16 = 7.38$ in.^{2}

To find the total area of the end view we must calculate the area of the fillets. The area of one fillet can be derived by subtracting one-fourth the area of the circle of the radius of the fillet of interest from one-fourth of the area of the smallest square that can contain the circle (Figure 6.20).

EXAMPLE
6.17

Solve for the area of one fillet in the end view of the part in Figure 6.19. Derive a direct equation to solve for the area of the four fillets.

Solution Referring to Figure 6.20,

area of fillet $= 1/4\, A_{\text{square}} - 1/4\, A_{\text{circle}}$

$A_{\text{fillet}} = 1/4\, (s^2 - \pi r^2)$

$A_{\text{fillet}} = 1/4\, (1.0^2 - \pi 0.5^2) = 0.0537$ in.^{2}

area of four fillets $=$ (area of square $-$ area of circle)

$A = s^2 - \pi r^2 = 1.0^2 - \pi \times 0.5^2 = 0.215$ in.^{2}

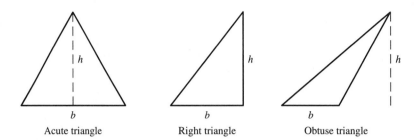

Acute triangle   Right triangle   Obtuse triangle

Figure 6.21 All of the general types of triangles shown above have an area equal to one-half the base times the height.

The total area of the end view is the area of the two flanges, the web, and the four fillets:

Total area $= A_{\text{total}} = 10.32 + 7.38 + 0.21 = 17.91$ in.^{2}

Once the total area of the end view is determined, we may solve for the volume. Because volume involves three-dimensional figures, we will continue the calculation for the weight of the part in the section on solid geometry.

Triangles

Next, consider the area of the triangle. The triangle represents another major classification of two-dimensional shapes. The area of *any triangle* may be calculated by the equation $A = 1/2\ bh$ (Figure 6.21).

The sawtooth waveform in Figure 6.22 provides a horizontally timed trace for an oscilloscope. The **average value,** which will equal the measurement of the voltage on a DC voltmeter, is equal to the area under the graphed outline of the function divided by the length of the function.

EXAMPLE 6.18

What would a DC voltmeter read when connected across the horizontal amplifier's output?

Solution The sawtooth waveform has a peak voltage, or height, of 35 V and is set by the time-base amplifier on the "scope" to occur every 5 ms.

The area A is

$$A = 1/2\ bh = (0.5) \times (5 \times 10^{-3}\ \text{s}) \times (35\ \text{V})$$
$$= 0.0875\ \text{volt} \cdot \text{seconds (V} \cdot \text{s)}$$

And the average value, as read by a DC voltmeter, is

$$V_{\text{average}} = \frac{0.0875\ \text{V} \cdot \cancel{s}}{0.005 \cdot \cancel{s}} = 17.5\ \text{V}$$

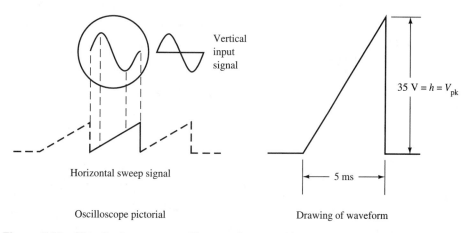

Oscilloscope pictorial

Drawing of waveform

Figure 6.22 The display on an oscilloscope is created by a sawtooth waveform applied to the horizontal plates inside the cathode-ray tube (CRT). The waveform moves the electron beam across the screen at regularly timed intervals.

We can conclude from Example 6.18 that any triangularly shaped electrical signal will have an equivalent DC value (average value) of one-half the maximum or peak voltage. The following example will show that any triangularly shaped graph will have an average value of one-half of the maximum value.

EXAMPLE 6.19

An automobile undergoes constant acceleration from 0 to 80 km/hr in 50 s, then a constant deceleration from the 80 km/hr to a final velocity of 0 in 5 s. How far did the car travel during the period?

Solution First, draw a graph of the triangle formed by the car's motion (speed vs. time). See Figure 6.23.

Next, solve for the area under the resultant triangle:

$A = 1/2\ bh = (0.5)(55\ \text{s})(80\ \text{km/hr}) = 2200\ \text{km} \cdot \text{s/hr}$

Then, solve for the average velocity by dividing by the total time, or the distance under the graphed curve:

$$V_{\text{average}} = \frac{2200\ \frac{\text{km} \cdot \cancel{s}}{\text{hr}}}{55\ \cancel{s}} = 40\ \text{km/hr}$$

and the distance traveled is computed by

$$\text{distance} = \text{velocity} \times \text{time} = \frac{40\ \text{km}}{1\ \cancel{\text{hr}}} \times \frac{1\ \cancel{\text{hr}}}{3600\ \cancel{s}} \times \frac{55\ \cancel{s}}{1} = 0.611\ \text{km}$$

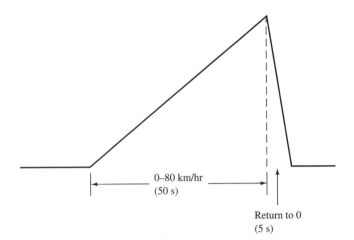

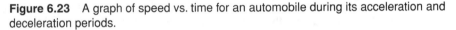

Figure 6.23 A graph of speed vs. time for an automobile during its acceleration and deceleration periods.

You should now be able to use the perimeter, circumference, and area equations with confidence. You will encounter geometry often in your technical career, and a firm understanding of plane and solid geometry will be of great benefit.

6.5 Solid Geometry for the Technologist

Three-dimensional shapes are studied under the discipline of solid geometry. We have already dealt with three-dimensional figures, the cone and frustum, in Examples 6.9 and 6.10. However, we used cross sections of the pieces to determine the angle of the cutting tool. We did not deal with the volume and surface area equations for these figures.

To illustrate surface area and volume concepts for solids, consider the right-circular cylinder in Figure 6.24A. The equation for the surface area of the cylinder is

$$A_{\text{surface}} = A_{\text{top and bottom}} + A_{\text{side}}$$

Because the areas of the top and bottom are equal to the area of a circle, and the area of the side is equal to the rectangle (see the cylinder unwrapped in Figure 6.24B), we can say

$$A_{\text{surface}} = 2(\pi r^2) + L \times W$$

and, because W is equal to the circumference of the top or bottom and L to the height,

$$A_{\text{surface}} = 2(\pi r^2) + 2\pi r \times h$$

After factoring out the common term of the polynomial, we find the surface area of a cylinder to be equal to

$$A_{\text{surface}} = 2\pi r(r + h)$$

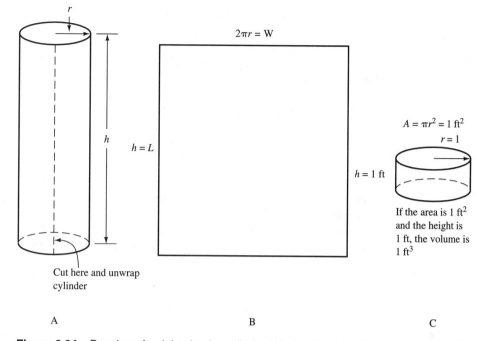

Figure 6.24 Drawing of a right circular cylinder (A), the circular cylinder unwrapped (B), and the concept drawing for calculating the volume of the cylinder (C).

The equation for the volume involves multiplying the area of one end (i.e., the top or bottom) times the height, as shown in Figure 6.24C.

$$V_{\text{cylinder}} = A_{\text{top or bottom}} \times \text{height} = \pi r^2 h$$

EXAMPLE 6.20

A cylindrical refinery tank is 35 ft high and 25 ft in diameter. Find (1) the amount of paint, in square feet of coverage, to paint the tank and (2) the amount of fuel oil it contains in gallons.

Solution To find the amount of paint, we must solve for the surface area by using

$$A_{\text{surface}} = 2\pi r(r + h)$$

$$A_{\text{surface}} = 2 \times \pi \times 12.5 \text{ ft} \times (12.5 \text{ ft} + 35 \text{ ft})$$

$$= 6.28 \times 12.5 \text{ ft} \times 47.5 \text{ ft} = 3730 \text{ ft}^2$$

Enough paint must be purchased to cover at least 3730 ft^2 of surface area. Note that the units, square feet, are appropriate for the area calculation.

The number of gallons the tank holds is a volume calculation. The information on the inside front cover shows a gallon to be equivalent to a volume of 231 in.^{3}. The volume of the cylinder is

$$V_{\text{cylinder}} = \pi \times r^2 \times h = 3.14 \times (12.5 \text{ ft})^2 \times 35 \text{ ft}$$

$$= 3.14 \times 156 \text{ ft}^2 \times 35 \text{ ft} = 17\ 200 \text{ ft}^3$$

The tank will hold 17 200 ft³ of liquid. By using the unit ratio method to convert to cubic inches and to gallons, we find the tank will hold

$$\frac{17,200 \text{ ft}^3}{1} \times \frac{12^3 \text{ in.}^3}{1^3 \text{ ft}^3} \times \frac{1 \text{ gal}}{231 \text{ in.}^3} = 129\ 000 \text{ gal}$$

If you have followed the development of volume presented in Chapter 4 and in the last example, you now know that the volume of a solid is the area of one end times the length, if the area is the same over the entire length. See Problems 41 and 42 for further applications.

We now return to the scrap problem we began in the last section (Figure 6.25). To calculate the value of the scrap or rework for one of these parts, the average weight

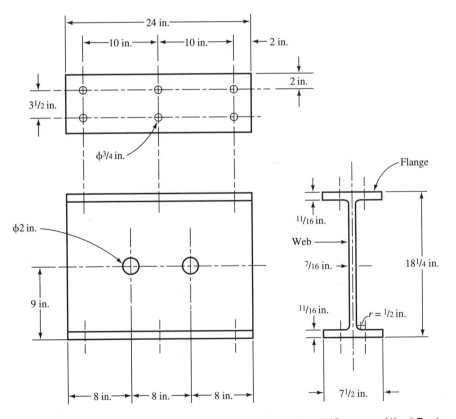

Figure 6.25 Full orthographic drawing of part for power shovel. (Courtesy Alfred Penko, Cuyahoga Community College)

density (0.280 lb/ft^3) of the material purchased must be multiplied by the volume to obtain the weight of each piece.

EXAMPLE 6.21

Complete the weight calculation for the part in Figure 6.25.

Solution   Referring to Example 6.16, the area of the end view has already been calculated as 17.91 in.^{2}. The area of the part is constant over its length (24 in.), so the volume may be calculated by multiplying the area times the length.

$$V = A \times L = 17.91 \text{ in.}^2 \times 24 \text{ in.} = 430 \text{ in.}^3$$

Next, we will solve for the volume of the holes by using the cylinder equations developed in this section. Because the diameters of the holes are given, not the radii, we will convert the area equation (πr^2) to an equation using the known diameter:

$$A_{\text{holes}} = \pi r^2$$

and, by converting radius to diameter,

$$\pi \frac{(d)^2}{(2)^2} = \frac{\pi}{4} d^2$$

The volume equation in terms of diameter becomes

$$V = \frac{\pi}{4} d^2 h$$

Therefore

$$V_{2 \text{ in. holes}} = \frac{\pi}{4} (2)^2 \, 0.438 = 1.38 \text{ in.}^3$$

Because there are two of these holes per piece, the total volume of the 2 in. holes is 2.76 in.^{3}.

$$V_{0.75 \text{ in. holes}} = \frac{\pi}{4} (0.75)^2 (11/16) = \frac{\pi}{4} (0.75)^2 \, 0.688 = 0.304 \text{ in.}^3$$

and, because there are twelve of these holes per piece, the total volume of the 3/4 in. holes is $12 \times 0.304 = 3.65$ in.^{3}

Adding the volumes of the 2 in. and 3/4 in. diameter holes, we find the total volume of the holes to be equal to

$$V_{\text{holes}} = 2.76 \text{ in.}^3 + 3.65 \text{ in.}^3 = 6.41 \text{ in.}^3$$

and, subtracting the volume of the holes from the total volume,

$$V = 430 \text{ in.}^3 - 6.41 \text{ in.}^3 = 424 \text{ in.}^3$$

Note: When rounding numbers to be added or subtracted, first perform the operation, and then round off to the decimal place accuracy of the least accurate figure.

We are finally ready to calculate the finished weight of the manufactured part. Because the weight density (0.280 lb/in.^{3}) is already in units of in.^{3} the multiplication can be performed directly as

$$\text{Weight in pounds} = \frac{424 \text{ in.}^3}{1} \times \frac{0.280 \text{ lb}}{1 \text{ in.}^3} = 119 \text{ lb}$$

In this example, the volume of the holes may not have been of importance. But the use of the geometry confirmed the limited effect on the final calculation. The example also may help you to solve more detailed problems in the future.

It is hoped this chapter has given you insight into how important visualization—coupled with an understanding of trigonometry and geometry—is to technical problem solving.

Problems

Section 6.1

1. Using the Pythagorean theorem, solve for the missing sides of the following right triangles. A decimal answer is acceptable.
   a. side $a = 6$, side $b = 8$, side $c =$
   b. $a = 0.04, b = 0.09, c =$
   c. $b = 19.2, c = 28.7, a =$
   d. $a = 1/4, c = 1, b =$
   e. $a = 12$ mi, $c = 13$ mi, $b =$ _____ mi

2. Solve for the missing sides of the following triangles. A decimal answer is acceptable.
   a. side $a = 2$, side $b = 9$, side $c =$
   b. $a = 0.07, b = 0.05, c =$
   c. $b = 28.9, c = 65.4, a =$
   d. $a = 1/32, c = 7/16, b =$
   e. $a = 844$ m, $c = 915$ m, $b =$ _____ m

3. For each of the following angles, find the complementary angle.
   a. 65°   **b.** 50.9°   **c.** 7.15°   **d.** 4.54°   **e.** 0.004 45°

4. See the sine, cosine, and tangent relationships written for angle A in the standard triangle (see Figure 6.4). Write the relationships again, but for angle B. Example:

   $$\sin B = \frac{\text{opp}}{\text{hyp}} = \frac{b}{c}$$

5. In Example 6.4 (also Table 6.1), explain why the sine and cosine functions are equal at 45° and why the tangent function is equal to unity at that angle.

6. In Example 6.4, explain why the tangent of 90° yields an error flag on the display.

7. Using your calculator solve for the trig functions (sine, cosine, and tangent) of the following angles (use three-significant-figure accuracy in your answers).
   a. 12°   **b.** 37°   **c.** 88°   *****d.** 0.012°

8. Use your calculator to solve for the trigonometric functions of the following angles.
   a. 24°   **b.** 63°   **c.** 43.2°   **d.** 89.7°   **e.** 12 600°

*Challenging problem.

9. Find the inverse sine ($\sin^{-1}$), cosine ($\cos^{-1}$), and tangent ($\tan^{-1}$) functions for the following quantities on the calculator.
   a. 0.5 **b.** 1.0 **c.** 1.38 (arc tangent only)
   d. 0.866 **e.** 0.707 **f.** 0.999

10. On your calculator find the inverse, or arc, sine, cosine, and tangent functions for the following ratios (of the sides of the right triangle).
   a. 0.800 **b.** 0.323 **c.** 4.47 (arc tangent only)
   d. 0 **e.** 0.700 **f.** 1

Section 6.2

11. Given the following information for a right triangle in standard position (see Figure 6.4), solve for the selected side or angle (draw the triangle for each problem before solving).
   a. $a = 3, c = 5, A =$ _____ °
   b. $a = 0.79, c = 1.88, B =$ _____ °
   c. $A = 37°, b = 18, c =$ _____
   d. $B = 12°, c = 85.3, b =$ _____
   e. $A = 53.9°, B =$ _____ °, $b = 0.003, a =$ _____

12. Solve for the unknown sides or angles for the following right triangles.
   a. $a = 6, c = 10, A =$ _____ °,
   b. $a = 0.23, c = 1.90, B =$ _____ °
   c. $A = 64°, b = 75.6, c =$ _____
   d. $B = 41°, c = 40.6, b =$ _____
   e. $A = 34.5°, B =$ _____ °, $b = 0.952, a =$ _____

13. Refer to Appendix C. Explain how the ancient village of SOH-CAH-TOA will help you to remember the three trigonometric relationships.

14. Solve each of the 3-4-5 triangles in Figure C.2 of Appendix C for the unknown illustrated by a wavy line. Show all solution steps for each problem.

15. See Example 6.7. Use the tangent function for angle A to solve for the percent grade [($\tan 8°$) $\times$ 100 = _____]. Is the tangent function's definition (the ratio of the opposite over the adjacent side) equivalent to the percent grade (the ratio of the rise over the run $\times$ 100)?

16. The percent grade of a slope is 48.4. If the thickness of the base is 43 ft, what is the height? (To solve, first draw a cross-sectional view.)

17. The guy wires for an antenna tower are attached 36 m above the ground. If the other end of the guy wires must be attached to the ground plane, 42 m from the base of the tower, to give appropriate stability, what is the length of each wire? What is the angle of the wires to the ground? What is the angle of the wires to the tower? (*Hint:* Draw a diagram and label all sides and angles before solving.)

18. A cone must be turned on a lathe. If the height of the cone is 25 cm and the diameter of the base is 65 cm, what is the angle of a cross section at the base (see Figure 6.10)? What will be the setting of the compound rest necessary to machine the cone?

19. See Example 6.10. Redraw the frustum to show how one of the triangular portions is dimensioned. Write the equation for solving for angle A in terms of the frustum height and base dimensions.

Section 6.3

20. Complete Table 6.2 by writing the missing equations to solve for the movements in the north and east directions.

*21. Using polar to rectangular conversion on your calculator, convert the vector 3000 at 21.5° to rectangular (X, Y coordinates). (The application is taken from Example 6.11.)

*22. Using rectangular to polar conversion on your calculator, convert the vector ($X = -2226$ and $Y = 2701$) to polar form ($R \angle \theta$). (The application is taken from Example 6.11.)

23. Solve for the vertical and horizontal components of the following vectors (all angles are defined from the horizontal):
   a. 358 at 45°
   b. 24.7 at 30°
   c. 6.90 at 88°
   d. 0.354 at 0.145°
   e. 3490 at 0°
   f. 1/8 at 28°

24. See Example 6.13. Using graph paper, construct the vector triangle to a scale of 1 mm = 100 lb of force, and solve for the resultant force vector ($R = 8890$ lb at 13° from vertical) by measurement.

25. In Figure 6.26, solve for the total vertical downward force on the beam.

26. Use the vector-triangle method to solve for the total force of the vectors in Figure 6.26.

27. See Example 6.14. Set up the unit conversion to solve for the angle of the voltage sine wave at 7 ms.

28. In Example 6.14, write the equation and solve for the instantaneous voltage (e) at $t = 13$ ms.

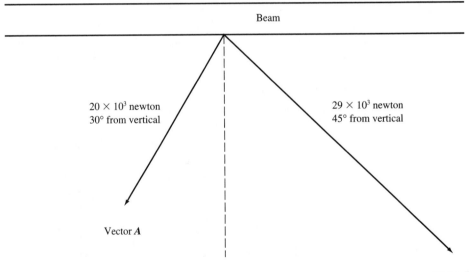

Beam

20 × 10³ newton
30° from vertical

29 × 10³ newton
45° from vertical

Vector **A**

Vector **B**

Figure 6.26 The two concurrent force vectors acting on a beam are to be resolved into (1) the vertical or downward components and (2) the total force component.

*Challenging problem.

Section 6.4

29. It is extremely difficult to measure the circumference of a glass. But if you invert the glass on a counter, wrap a string around it, and carefully mark the intersection points on the string, it can be done to a millimeter of accuracy. Measure the diameter of the glass in millimeters and divide. With an accurate procedure you can divide the diameter into the circumference and obtain a result of 3.1 ± 0.1, a value that approximates the unitless ratio π.

30. Solve for the circumference (c) and area (A) of circles with the following *diameters:*
    a. 861 **b.** 81.8 **c.** 0.828 **d.** 12 mi **e.** 63 km

*31. Refer to Figure 6.20. Show by a drawing, and the area equations for a circle and a square, how the area of four fillets is equal to $0.215 \, s^2$, where s is one side of the square.

32. The triangular section of a large parking lot must be asphalted. If one side is 120 m and the depth or altitude of the triangular section is 87 m, what is the area of the lot?

*33. Using graph paper, construct a drawing like that of the similar triangles for $45°$ in Figure 6.5. Change the base angle to $30°$ and use 30, 60, and 90 mm as the length of the bases b', b'', $b' '''$. Prove by measurement that the ratios of the sides of the triangle agree with the sine, cosine, and tangent functions computed on your calculator.

*34. Make the drawing constructed in the previous problem into a rectangle by adding a top (horizontal) line as long as the base and a left-hand side (vertical) line as long as the altitude. Solve for the area of the resultant rectangle. What is the area of the triangle? Is this problem-solving process the same as using the equation $A = 1/2 \, bh$?

35. A swimming pool has a depth of 4 feet at the shallow end and 10 feet at the deep end. See Figure 6.27 for further dimensional information on one of the long sides of the pool. Solve for the area of the side at the shallow end of the pool (the rectangular area).

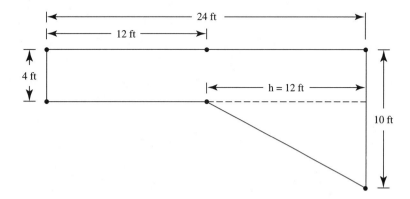

Figure 6.27 One of the long sides of a swimming pool. *h* refers to the height of the trapezoid at the deep end of the pool.

*Challenging problem.

36. In Problem 35, solve for the area of the side at the deep end of the pool. Refer to Figure 6.18, the *trapezoid.*

37. A sawtooth waveform (see Example 6.18) has a peak voltage of 5 V. What is the voltage read by a DC voltmeter?

38. Solve for the average velocity of an automobile that accelerates from 0 to 60 mph in 22 s and decelerates, immediately, to 0 mph in 18 s (see Example 6.19).

Section 6.5

39. Calculate the volume and surface area of the following right-circular cylinders:
    a. a .22 caliber shell casing that measures 0.22 in. in diameter by 1.0 in. in length
    b. a storage tank measuring 40 ft in diameter by 20 ft in height
    c. a storage tank measuring 30 m in diameter by 25 m in height

40. Determine the volume of the two storage tanks in Problem 37 in terms of gallons for the storage tank in **b,** and in liters for the storage tank in **c.**

41. Many storage tanks are hollow spheres. The volume of a sphere is solved for by the following equation: $V = 4/3 \pi r^3$, where r is the radius of the sphere. Solve for the volume of the hollow portion of the sphere in Figure 6.28.

*42. In Figure 6.28, solve for the volume of the material that makes up the walls of the hollow sphere. You must use the information solved for in Problem 41.

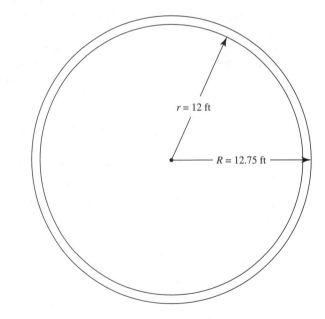

$r = 12$ ft

$R = 12.75$ ft

Figure 6.28 A hollow sphere. Drawing is not to scale.

*Challenging problem.

43. A standard $W21 \times 142$ I-beam has the following dimensions (Figure 6.27): $d = 21.46$ in., $t_f = 1.095$ in., $t_w = 0.659$ in., and $W_f = 13.132$ in. Solve for the cross-sectional area of the beam, ignoring the area of the fillets.

All Sections

For problems 44 to 47, first make a sketch of the problem. Perform the necessary measurements in metric or English system, then convert to the other system and offer a second solution.

44. Estimate the height of a large tree, by pacing off the estimated distance (approximately 3 ft per step) from the tree's base, and by measuring the angle from the local horizon to the top of the tree (use a make-shift plumb bob attached to the small hole on the base of your protractor).

45. Estimate the area of a driveway or parking lot by assigning each section of the lot an appropriate geometry (see Problem 32).

46. Find the height of a point on your campus as referenced to a benchmark. Your instructor will provide equipment and further information.

***47.** Solve for the volume and weight of a part by estimating certain geometries on the part (as in Examples 6.16 and 6.21). Find the weight density of the part's material by using look-up tables in an engineering handbook, or by experiment. To experiment weigh the part in air and then as it is submerged in water. Use the following equation to solve for the weight density:

$$\text{weight density} = \frac{\text{weight in air}}{(\text{weight in air} - \text{weight in water})} \times 62.4 \, \frac{\text{lb}}{\text{ft}^3} \, (\text{weight density of water})$$

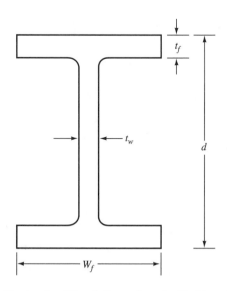

Figure 6.29   I-beam with standard literal dimensions for Problem 43.

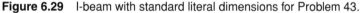

*Challenging problem.

Selected Readings

Harris, Norman C., Edwin M. Hemmerling, and A. James Mallmann. *Physics: Principles and Applications.* 5th ed. New York: McGraw-Hill, 1990.

Jacobs, Harold R. *Geometry.* San Francisco: W. H. Freeman and Company, 1974.

Kramer, Arthur D. *Fundamentals of Technical Mathematics with Calculus.* 2nd edition. New York: McGraw-Hill, 1989.

Oberg, Eric, with Franklin D. Jones, and Holbrook L. Horton. *Machinery's Handbook.* 25th ed. New York: Industrial Press, 1996.

7

The Technical Laboratory

The earliest technicians were laboratory technicians. They were recruited from industrial jobs because of their ability to understand processes and often for their unusual knowledge and interest in the math-sciences. These "super-techs" worked side by side with scientists to perform the vital research and development (R & D) of industry.

Many of you will be continuing that important work. Some of you will work in research laboratories developing the latest products (Figure 7.1). Others will solve current industrial-process problems by using quality control laboratories. Still others will be required to experiment with industrial processes on the factory floor, in effect making the workplace a laboratory. It is no wonder, then, that the laboratory is the heart of a technical education; it provides the critical hands-on experience portion of the curriculum. *This chapter is intended to make you aware of how to work safely and effectively in the laboratory and report your findings clearly and efficiently.*

Most technical colleges have invested in costly industrial-grade equipment, so you will be most productive at graduation even with little or no work experience. Of course, you will be expected to treat the equipment with care.

Because much of the equipment in the technical laboratory can be dangerous, you will be instructed how to *work safely* with it. Safety considerations learned in the technical laboratory could save your life—especially later, while on the job. Always listen carefully to your instructor when in the lab. Take all safety rules seriously. Consider the consequences of not following suggested practice, and do not be afraid to ask questions when in doubt of the appropriateness of a rule or practice.

The occupational satisfaction questions raised in Chapter 1 all apply to the laboratory; therefore, laboratory activities can and will be effective in helping you decide for or against a technical career.

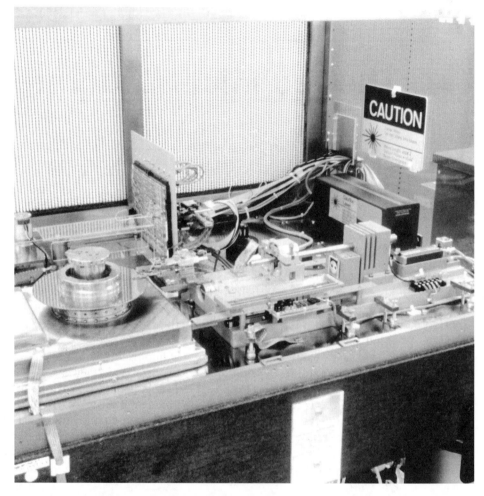

Figure 7.1 A precision test stand used in R & D of computer disks. (Courtesy of International Business Machines Corporation. Unauthorized use not permitted.)

To be a good laboratory technician, you must

1. **Be patient and pay close attention to detail.** Methodically performing the steps of an experiment may sometimes be frustrating. Many times the experiment does not turn out as planned; this means it may have to be redone, costing much precious time and energy.
2. **Be a good planner,** have confidence in your plan, and prepare well to accomplish the main objective of the assigned lab. Often, your instructors will be willing to go only part of the way in explaining theory and methodology, not because they do not want to help, but because the greatest benefit will come from figuring it out for yourself.

3. **Be accurate and objective in data collection.** Simply recording data expected, or wished for, from the experiment and not mastering the skills of accurately and objectively interpreting meters, gauges, or other measurement devices will eventually be detected on a report or on a lab practical examination.
4. **Communicate clearly and effectively.** Only when the reader can understand what you are trying to convey is your work of any benefit.

Underlying all of the material in this chapter is another way of thinking about problems and solving them. (A general problem-solving method was described in Section 3.3, *Problem Solving.*) This more specific method of problem solving is used in the technical laboratory and is known as the *scientific method.* The five steps of the scientific method are described and illustrated below:

- *State the Problem.* When working in the laboratory the technician should be able to identify the problem to be solved. (There is always a problem to be solved—laboratories are far too expensive to allow work that is not based on a specific problem.) For example, your company manufactures a container (a box) that is to be used to ship computers. Your specific work is to drop-test the computer packed in the box. You state the problem as "Can the computer survive the drop test?"
- *Form the Hypothesis.* A hypothesis is a stated assumption or set of assumptions used in the experiment. For the example of the drop test, the assumption is that the experiment fairly models the actual stresses the computer will experience during shipping.
- *Experiment and Observe.* You test a computer before it is packed in the box. You assure it is packed as intended and you perform the drop test. You unpack the computer and retest for any damage. During the entire procedure you must guard against errors in the test. For instance, was the equipment used for the drop test set at the correct height?
- *Interpret the Data.* In this case, the computer passed the retest and no visible damage was noted. A check sheet would be used at this point to determine if the retest covered all possible problems or malfunctions. The visual inspection check sheet might have a drawing of all sides of the computer for you to mark the position of any dents or abrasions.
- *The Conclusion.* Your conclusion should be stated clearly and concisely for each experiment. In this case, the conclusion is that the container protected the computer in this particular drop test.

7.1 Performing the Experiment

For the aspiring technologist, the greatest learning will take place in the laboratory. You will remember many of the hands-on experiences the laboratory provides for the rest of your life. Whether you are in a general laboratory, such as physics, or in a technology-specific laboratory, many or most of the labs will be spelled out in some detail.

Equipment will be specified and documented. Prepared data sheets will be offered. Often, the theoretical concepts involved in the experiment will be described. Even with such support information, experimentation will take you into the unknown because there is great variability in technique, methodology, environmental conditions, and even in the inherent characteristics of the equipment, such as friction. Identifying the reasons for variability is one of the great lessons you will learn in experimentation—no two experiments are the same.

In a few cases, especially after entering the second year of study, only the experimental "statement of the problem" to be explored is described. For instance: "The student will design an amplifier to convert a 5 millivolt (mV) signal to a 10 V signal," or "The student will determine the bulk modulus and compressibility of a concrete sample." You must make decisions on appropriate equipment, review equipment manuals for instrument accuracy and precision, and perhaps read research reports on ex-

Figure 7.2   You are responsible for results. (Courtesy H & N Instruments)

periments similar to what you are attempting. Your skill, creativity, and perseverance will really be challenged. But these laboratories will prepare you for the real world of industry, where failures occur frequently. Good technicians learn from these failures, and new, improved materials and systems result.

Often, you will work with others on group projects, just as you will be solving problems with others in industry. Always keep in mind, however, that you alone are responsible for the final results. It is ultimately up to you to verify data, assure correct equipment setup, and draw conclusions from the results of your laboratory experience (Figure 7.2). See Chapter 3's problem-solving example (Example 3.1) for further discussion of this important, individual aspect of the laboratory.

7.2 **Reducing Errors in the Laboratory**

During the experiment you will encounter variability. All of your measurements are subject to variability, and you should learn to recognize the major causes (some experimenters refer to major causes of variability as the *red X*s). You must eliminate the red *X*s if possible and record those impossible to eliminate. Later, when you write your report, you will be able to explain the causes of error that occurred during the experiment.

To better understand errors it is best to classify them in three general categories: gross, systematic, and random. As you read the remainder of this section, refer to Figure 7.7 (p. 200).

Gross errors are errors that completely invalidate data. Of course, these errors should be eliminated in the laboratory and *not be discussed in the report*. Gross errors include misuse of equipment, not following the procedure or the proper sequence of the procedure, or recording data incorrectly.

Systematic errors are errors exhibiting an orderly character. Systematic errors may be caused by the environment, the measuring instrument, or the experimenter. Systematic errors may also be identified and eliminated during the laboratory and often will not show up in a report. Figure 7.3 is an example of what could be considered either a gross or systematic error.

Not zeroing the meter is a gross error because it represents misuse of equipment. Zero offset may also cause a systematic error, but in the case of an ohmmeter the error may not be systematic. You cannot assume that zero offset is always a systematic error.

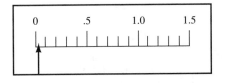

Figure 7.3 An indicator not zero-set usually causes a systematic error (a consistent error). However, in some cases it will cause random errors (e.g., ohmmeter errors).

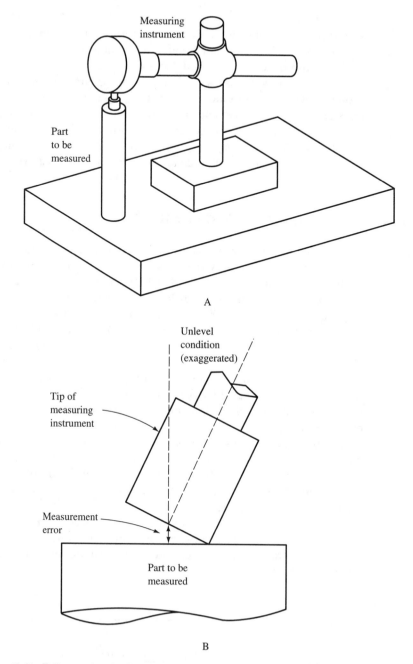

Figure 7.4 Failure to level a height gauge could result in a systematic error known as cosine error. In B, the error can be readily seen. Can you see why it is referred to as cosine error?

Examples of systematic errors are equipment out of calibration, failure of mechanical devices, such as a steel measuring tape that lengthens with temperature change (ambient temperature should often be monitored during experimentation), gauges affected by barometric pressure, carbon resistors that tend to increase in resistance as they are used over and over again in the laboratory, and oscilloscope sensitivity adjustments that are not locked into the calibrate (cal) mode. Your instructor will no doubt demonstrate typical systematic errors you will encounter in specific laboratories. See Figures 7.4A and 7.4B for another example of a systematic error.

Some instruments display a systematic error known as hysteresis. Hysteresis is a Greek word meaning *lag*. A more complete definition for **hysteresis** is "the inability of a moving or deformed object to follow the same displacement curve in both increasing and decreasing directions" (Figure 7.5). Some laboratory or instrument user's manuals will ask the technician to always approach the final setting from one direction only—e.g., clockwise or counterclockwise.

Observation errors are often systematic. **Observation errors** are errors caused by an observer misreading the scale of an instrument. The most common type of observation error is parallax. **Parallax** is an error resulting from the indicating needle on an analog scale being some distance from the scale (Figures 7.6A and 7.6B). A technician who consistently reads the instrument from the side will have a systematic error in the data collected.

Another frequent error is adjusting the data to fit the desired or theoretical results. This may not be deliberate on the part of the experimenter, but may simply result

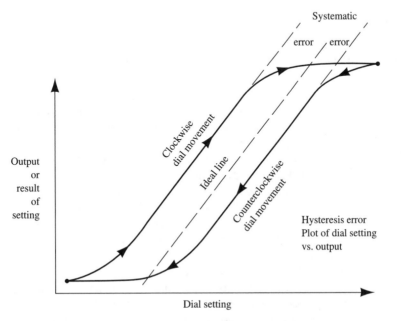

Figure 7.5 A typical hysteresis curve, usually attributed to mechanical "lag" in an instrument.

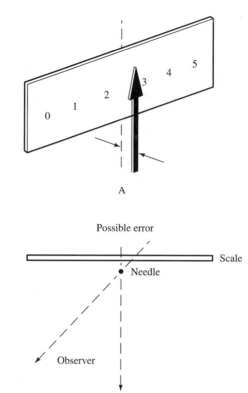

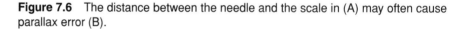

Figure 7.6 The distance between the needle and the scale in (A) may often cause parallax error (B).

from taking readings several times before recording a particular data point. Because there is variability in all experiments, you can eventually obtain a reading close to the one you want or expect. Inspectors in industry are often guilty of "editing the data" so product quality will look good. Editing the data or biasing the data is also considered systematic in that all data will be closer to the ideal than they would if the experimenter recorded every measurement—not just good measurements. Beware of such conscious or unconscious activity. It will lead to unpredictable or unreproducible results.

Random errors are errors that are due to chance causes, often affecting the measurement at different times and in varying directions. Friction is a good example of a random error. Friction may always be in one direction (for instance, in an inclined plane experiment), but the magnitude of the error will usually vary, or never be constant.

Environmental errors have already been mentioned as a type of systematic error. Many environmental errors are random. Drafts or air currents often disturb the measurement system. Unknown temperature fluctuations cause the components of sensitive measurement equipment to expand and contract in an unpredictable fashion. Relative humidity (the amount of moisture in the air) may reduce static electricity forces, but increase small friction—"stiction"—forces. Vibration is also considered an environmental error. Sometimes vibration is purposely introduced at a pivot point to eliminate stiction.

Random errors can never be eliminated, but the impact of such errors can be reduced by the use of statistics. Averaging data is a simple way to reduce the effects of random errors.

EXAMPLE 7.1

Five pieces from an assembly line have had one quality characteristic, designed to be 5.33 mm, measured by a micrometer as follows: 5.34 mm, 5.30 mm, 5.34 mm, 5.33 mm, and 5.31 mm. Assuming the variability to be due to chance or random variation, find the true average or mean ($\overline{X}$) of the manufacturing process and determine if the process is centered.

Solution The average, mean, or $\overline{X}$ is calculated by

$$\overline{X} = \frac{5.34 + 5.30 + 5.34 + 5.36 + 5.31}{5} = 5.33 \text{ mm}$$

The process is centered and the operator may continue to produce parts, not stopping to adjust the process.

In the next section, "Data Collection and Calculating Results," you will learn that answers to a calculation should be rounded to the number of significant figures of the number with the least number of significant figures. *The 5 in this example is a counting number and not a data number.* The data are accurate to three significant figures, so the answer is accurate to three significant figures.

Percent error (called *relative error* in some textbooks) is the combined error of the experiment. It may be determined by comparing the value of a parameter determined by experiment to the theoretical value (accepted true value) of the parameter under study. Put into equation form:

$$\text{percent (\%) error} = \frac{|\text{experimental value} - \text{accepted true value}|}{\text{accepted true value}} \times 100$$

For the measured values in Example 7.1, the percent errors from the $\overline{X}$ are calculated as shown in Table 7.1.

Table 7.1 Percent Error Calculations for Example 7.1*

Measurement	% Error Calculation*	% Error
5.34	$|5.34 - 5.33|/5.33 = 0.0019$	0.19
5.30	$|5.30 - 5.33|/5.33 = 0.0056$	0.56
5.34	repeat calculation	0.19
5.36	$|5.36 - 5.33|/5.33 = 0.0056$	0.56
5.31	$|5.31 - 5.33|/5.33 = 0.0038$	0.38

*Note the use of the absolute value sign and the difference of the numerator is always positive. All values are dimensionless; 5.33 is the theoretical or expected outcome—the *accepted true value*—of the experiment.

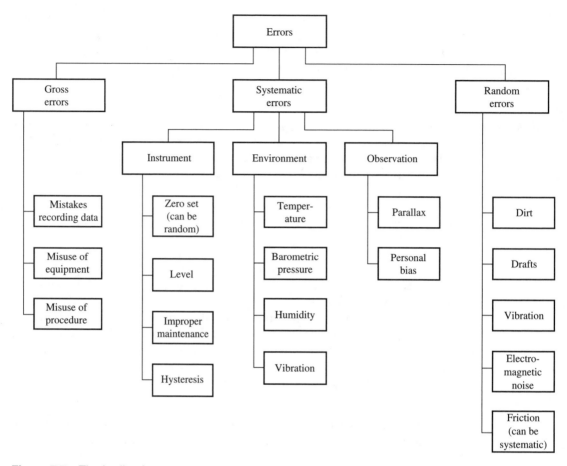

Figure 7.7 The family of errors.

As you can see, the **family of errors** (Figure 7.7) is large, and pinpointing a particular error is often difficult. Throughout your career you will be dealing with variability. To recognize and attempt to categorize the causes of errors will allow you to separate the natural variability of a manufacturing process or research experiment from variability caused by errors, gross or systematic, that can be controlled to improve the quality of the product.

The following definitions will aid you in developing a suitable vocabulary for explaining instrumentation practices in the laboratory.

ACCURACY: A quantitative estimate of the error or uncertainty in a measurement or calculation. The greater the accuracy, the smaller the value of the error.

CORRECTION: A quantity, equal to but opposite in sign from the error value, added to the measured value to obtain the accepted true value. Calibration often results in a correction factor for an instrument.

ERROR: The deviation of a calculated or observed value from the true or ideal value. Usually expressed as a percentage of deviation (percent error).

PRECISION: The ability of an instrument to repeat the same measurement after a certain length of time and often after varying the instrument value before resetting and performing the next measurement. Also, the instrument's ability to repeat a measurement and obtain the same result with a different operator.

RESOLUTION: A measure of how well a device can respond to small changes and indicate that change on the scale. Also referred to as sensitivity.

UNCERTAINTY: The sum of all of the errors influencing a measurement or calculation.

7.3 Data Collection and Calculating Results

Accurate reading of instrument scales is a challenge. First, the technician must determine what the scale divisions mean. In Figure 7.8, the smallest instrument scale graduations are read as A, 1.0; B, 0.2; and C, 0.1. These scale readings may then be multiplied by the scale multiplier.

It is important to understand how closely you can interpret a scale. The most common practice is *to consider the scale resolution to be no better than one-half of the smallest division.* The scales in Figure 7.8 can be interpreted to 0.5, 0.1, and 0.05 units, respectively. *Never* read (or interpret) an instrument scale to less than one-half of the smallest division. This number is known as the **doubtful digit.**

Most instrument manufacturers graduate an instrument's scale to approximate the accuracy of the instrument. **Accuracy** is defined as how true the value measured is. Only through traceability (see Chapter 5) can accuracy be determined, and because the manufacturer of the instrument may not guarantee accuracy to the smallest graduation, a plus and minus ($\pm$) value is stated in the documentation or owner's manual.

Precision also contributes to collecting "good" laboratory data. **Precision** is defined as the ability of an instrument to repeat a measurement (repeatability). Of course, the technician using the instrument affects the precision of the reading, especially when using an **analog** (as opposed to a digital) scale. Precision and accuracy are often confused. Figure 7.9 will help you separate the concepts of precision and accuracy.

Precision and accuracy may be reported as (1) percent of full scale, (2) $\pm$ some portion of the last digit, or (3) a combination of 1 and 2. See Figures 7.10A and 7.10B. A stated accuracy of ± 5 percent of full scale means that any reading taken with the scale in Figure 7.10A may be in error as much as 0.5 (full scale of 10 multiplied by 0.05).

EXAMPLE 7.2

The analog scale in Figure 7.10A is the indicator for a voltmeter with a stated accuracy of ± 5 percent. The reading is 13 V on the 100 V scale (full scale). Write the value of the reading and include the possible range of the measurement due to its stated accuracy.

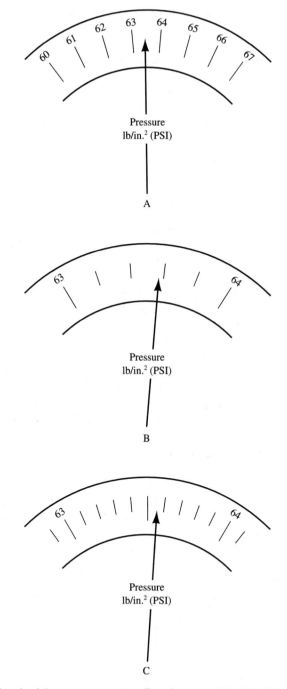

Figure 7.8 A sketch of three meter scales. Readings are (A) 63.5, (B) 63.6, (C) 63.55.

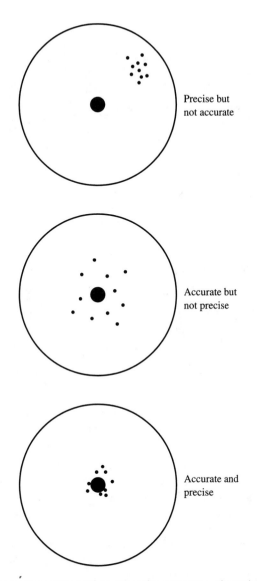

Precise but
not accurate

Accurate but
not precise

Accurate and
precise

Figure 7.9 These targets portray and contrast the concepts of precision and accuracy.

Solution Simply take 5 percent of the 100 V full scale,

$$0.05 \times 100 = 5 \text{ V}$$

and write the reading as 13 ± 5 V; the true voltage can be anywhere from 8 to 18 V.

Figure 7.10 (A) A 0 to 50 V and 0 to 100 V scale with ±5 percent error. (B) A digital display with a possible error of ±1 digit at the least significant digit.

EXAMPLE 7.3

The technician in Example 7.2 switches the voltmeter scale multiplier switch down to the 50 V scale (50 V is now full scale). Determine the new position of the pointer and calculate the range of the true value of 13 V read on this scale.

Solution The meter's pointer is now on the 13th scale division. Multiply 0.05 times 50, and write the true voltage as 13 ± 2.5 V. The true voltage "window" has been halved by moving to a lower, more sensitive scale.

EXAMPLE 7.4

For the digital display in Figure 7.10B, what is the confidence range of the reading?

Solution Refer to the caption for Figure 7.10B. The actual value of the reading could be anywhere between 873.3 and 873.5.

Most digital meters use a combination of accuracies. A typical accuracy may be read as "Accuracy is specified as ±([percent of reading] + [number of units in least significant digit])." Can you calculate the accuracy for the digital scale in Figure 7.10B if the stated accuracy is ±(0.5 + 2)? Use a scale reading of 1000 and see Problem 11 at the end of this chapter.

All data you collect in the laboratory are known as **primary data.** Technicians often make use of **secondary data:** that is, data that have been taken by someone else, and possibly weeks, months, or years ago. You may be directing operators to collect secondary data. More importantly, you will have to interpret these data later. Therefore, you need to specify the number of significant figures in which the data will be recorded, or chaos will result.

As an example, consider the scale in Figure 7.8A. The measurement is read as 63.5 PSI and the doubtful digit is 0.5. The 0.5 in the reading is underlined because it is estimated and considered to be a doubtful digit. If operators, as they were recording data, attempted to read the scale closer they would be beyond the accuracy of the instrument. Also, different operators may record the data to a different number of significant figures. You would have to direct the operators to record data to only three significant figures. If a pressure gauge with smaller scale graduations is used, as in Figures 7.8B and 7.8C, a more precise measurement can be made.

EXAMPLE 7.5

Read and record the pressure indicated in Figure 7.11 and underline the doubtful digit. To how many significant figures would you instruct operators to record data using this scale?

Solution   Use the previously stated guideline to interpret the scale to one-half of the smallest division. Since the pointer is centered between 63.5 and 63.6, the reading is interpreted as 63.55. The reading, with the doubtful digit underlined, is 63.55 PSI. Operators using this pressure gauge should record data to four significant figures of accuracy. If the pointer was only slightly to the right of 63.5, the reading would be 63.50. Never try to interpret a scale to less than one-half of the smallest division.

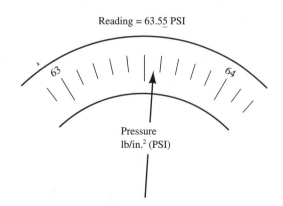

Figure 7.11   A pressure gauge with "unity" scale divisions.

Example 4.3 first discussed the concept of *significant figures*. For further clarification, the number of significant figures is defined as "a number equal to the number of digits the datum contains, with zeros merely holding a decimal place not counted."

Use the following rules for counting the number of significant figures in a quantity (the significant figures illustrating the rule are underlined):

1. All digits other than zero are counted as significant figures (example: 64.5).
2. All zeros between significant figures are also significant (example: 150.02).
3. For nondecimal numbers greater than one, all zeros placed after the significant figures are *not significant* (example: 500 000 or 5×10^5 [only the 5 is significant]).
4. If a decimal point is shown after a nondecimal number larger than one, the zeros are then considered significant (example: 500 000.).
5. Zeros placed after a decimal point that are *not* necessary to set the decimal point are significant (example: 500.00).
6. For numbers smaller than one, all zeros placed before the significant figures are *not significant* (example: 0.002 05 or 2.05×10^{-3} [only the three nonzero numbers (205) are significant; the other zeros establish the decimal point or fix the magnitude of the number]).

EXAMPLE 7.6	Determine the number of significant figures in the following quantities. Underline the significant figures.

Solution

1. 5100		has two significant figures
2. 2053		has four significant figures
3. 2 034 000.		has seven significant figures (because of the appearance of the decimal point)
4. 2 034 000.00		has nine significant figures
5. 0.000 032 1		has three significant figures
6. 0.000 032 01		has four significant figures

Once you have collected data and can appreciate the accuracy of the data taken you will begin to appreciate how to treat the result of multiplying and dividing such data.

With the advent of handheld calculators, students often report calculations to eight or ten significant figures. In other words, they record all or most of the numbers appearing on their particular calculator's display. Such a practice is very misleading— no quantity can be more exact or accurate than the numbers used to generate that

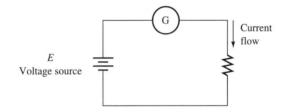

Figure 7.12 The sensitive galvanometer is connected in series with the resistor to measure current flow to an accuracy of 100 μA.

quantity. Consider the following example of the problems encountered when you ignore accuracy in terms of number of significant figures.

EXAMPLE
7.7

In the schematic diagram of Figure 7.12 the resistor has been measured in the laboratory as 24 Ω and the current is measured with a sensitive galvanometer as 3.658 2 A. The two numbers must be multiplied to determine the voltage ($E = IR$, Ohm's law). Solve for the voltage.

Solution

$$E = I \times R$$
$$= 3.658\ 2 \times 24 = 87.796\ 8 \text{ V}$$

But the 4 in 2$\underline{4}$ is doubtful. This means the true value of the resistor, according to the previous understandings of what doubtful digit means, could be between 23.6 and 24.4 Ω. What does this mean for our voltage calculation? Performing the multiplication again using the two extreme values,

$$23.6 \times 3.658\ 2 = 86.33$$
$$24.4 \times 3.658\ 2 = 89.26$$

we see that the true voltage is in *a range of almost 3 V.* This is to say that the accuracy of the procedure is ±1.5 V.

Clearly, the two results in the example vary so much that to report the data to more than the number of significant figures of the 24 Ω is inappropriate. Such reporting implies a precision of measurement greater than exists. To avoid reporting results that are misleading, use the following rule:

When multiplying or dividing quantities, record the result to the number of significant figures of the measurement with the *least* number of significant figures.

In Example 7.7, the resistor's value had two significant figures. Underlining the doubtful digits yields

$$2\underline{4} \times 3.658\,\underline{2} = 8\underline{7}.796\,8$$

We can see that the result should not contain more than two significant figures, and the next step is to round the result off to two significant figures:

$$2\underline{4} \times 3.658\,\underline{2} = \underline{88}$$

Consider the accuracy of the measurement to be $88 \pm 1.5 = 86.5$ to 89.5 V.

EXAMPLE 7.8   The following calculations further demonstrate the multiplication/division rule outlined above. Please confirm your understanding of each "rounding off" decision.

Solution

$$\underline{5}000 \times 3.25 = 16\,250, \text{ but is recorded as } \underline{2}0\,000$$

$$2010 \times 6.7\underline{7} = 13\,607.7, \text{ but is recorded as } 13\,\underline{6}00$$

$$\frac{1.025}{0.040\underline{2}} = 25.497\,512, \text{ but is recorded as } 25.\underline{5}$$

The multiplication and division rules do not apply for **counting numbers.** For example, when you average data you divide by the number of data points. The number of data points is not considered a datum, so it can be ignored in terms of significant figures.

When rounding a number to the doubtful digit use the following rules:

1. If the number to the right of the doubtful digit is greater than 5, increase the number by one:

   1$\underline{2}$.63 rounds up to 13

   363.$\underline{5}$7 rounds up to 363.6

   499 9$\underline{9}$9 rounds up to 500 000 (Does this example defy common sense?)

2. If the number to the right of the doubtful digit is less than 5, do not change the number:

   7.5$\underline{2}$47 rounds down to 7.52

   0.003 654 $\underline{9}$3 rounds down to 0.003 654 9

You have surely encountered these two rules before. If you change rule number 1 to include 5 (numbers *greater than or equal to 5* round up) they are all you need to handle data sets of less than 100. Later, you may collect data by computer for a manufacturing operation; manufacturing processes often yield large amounts of data. The following rule is used only when analyzing large amounts of data and ensures an equal distribution of "rounding ups" and "rounding downs."

3. If the number to the right of the doubtful digit is a 5, leave the number unchanged if it is even and round up if it is odd (i.e., always leave the doubtful digit an even number).

The rule for adding and subtracting numbers, when consideration of the number of significant figures is required, is quite different from the rule for multiplication and division. This is due to the nature of addition and subtraction. The decimal point must be aligned when adding and subtracting numbers (multiplication and division operations do not require decimal point alignment). Addition and subtraction operations, then, require a decimal place approach:

When adding or subtracting quantities, record the result to the number of significant figures of the number with the *least* number of digits beyond the decimal point. For nondecimal numbers greater than one, record the result to the last accurate number's place value (zeros placed after the last significant figure are not significant).

EXAMPLE 7.9

For the following calculations demonstrate the addition/subtraction rule above. Please verify all rounding off decisions.

Solution Unlike the multiplication/division rule, the decimal place rule may yield more significant figures than the number of significant figures of the individual elements.

3.25 (only three significant figures)
2.63
+ 4.25
10.13 (all four figures are significant)

12 523.544	12 523.544
– 200	– 200.
12 323.544, but is recorded as 12 300	12 323.544, but is recorded as 12 324 (Can you see why the two example answers differ?)

Before leaving this section, you should realize that the significant-figure approach, accuracy, and precision are complex, often interrelated, phenomena (also refer to Section 5.6). You will acquire further information regarding the many rules for handling data during your college education and in training programs on the job. Quality product and predictable manufacturing cannot be attained without accurate and reliable data collection methods.

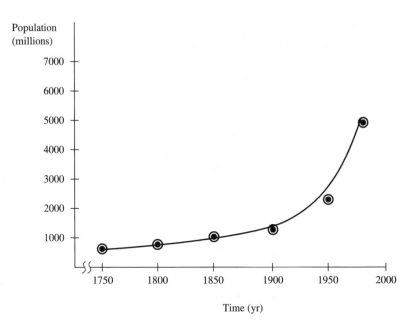

Figure 7.13 A graph depicting the change in population of the world with respect to time. Time is independent of population change, hence is the independent variable, and belongs on the *X*-axis.

7.4 Presenting Data—Graphing

After data are collected, they must be presented in a meaningful manner. In other words, they must be packaged in "consumable form." One way to present data, and the most frequent method used in the engineering technologies, is the line graph. A line graph displays the relationship that exists between two quantities. This visual representation is appealing to anyone working with numerical data—graphs show trends, discontinuities, and other characteristics that a list of numbers cannot show. Graphs assist the reader in understanding a relationship: "A picture is worth a thousand words."

The data you collect in the laboratory will often be the result of changing one variable and recording the change in another. It is hoped all other possible variables will be eliminated (e.g., friction in an inclined-plane experiment) or held constant. This is what is meant by a "controlled experiment."

The variable you control is called the **independent variable.** The data for this variable are always placed on the *X*-axis (horizontal axis, or abscissa; i.e., time in Figure 7.13).

The other variable is called the **dependent variable** (i.e., population in Figure 7.13) because its change is dependent on the change of the independent variable. If the two quantities under study are not related, a graph will show no relationship

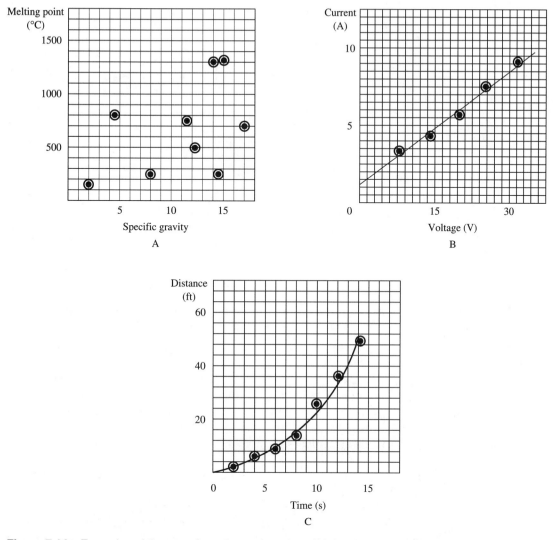

Figure 7.14 Examples of the use of graph paper to show (A) the change in melting point vs. specific gravity (these quantities are apparently not related), (B) current vs. the voltage drop across a resistor (these quantities are linearly related), and (C) the change in distance vs. time with constant acceleration (these quantities are exponentially related).

(Figure 7.14A). Usually, we experiment only with variables that obviously depend on the independent variable. The graphs in Figures 7.14B and 7.14C show variables that are related.

There are a number of points you must consider in laying out a graph. Only practice will benefit you here, but the following checklist will aid you in proofing your final product.

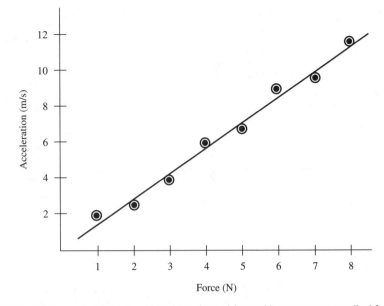

Figure 7.15 This graph of the acceleration of an object with respect to applied force demonstrates the appropriateness of the first five graphing guidelines.

1. Each axis must be properly labeled. Both quantity and units must be present (e.g., time, milliseconds [ms], or stress, newton/meter^2 [N/m^2]).
2. Each axis must be properly scaled. The axis should be clearly broken where a discontinuity exists (see Figure 7.13). Scale the axis using regularly spaced increments. Never use experimental point values exclusively in scaling the axis.
3. The independent variable should be on the X-axis and the dependent variable on the Y-axis.
4. Experimental points are plotted boldly (often circled).
5. A curve (a straight line is a curve of infinite radius) of best fit through the experimental points is drawn. Do not connect the dots! The graph in Figure 7.15 depicts the first five guidelines.
6. Analyze the curve to see if it fits the theoretical function, or the expected relationship (examples follow and your ability to do this will be augmented in your mathematics courses).
7. Title your graph after completion, when you can best establish the full objective of the illustration.
8. The finalized graph is inspected for accuracy and neatness.

Sometimes it is difficult to determine which variable is independent. In Figure 7.16, both graphs are of fuel efficiency (consumption) in gallons/mile as related to

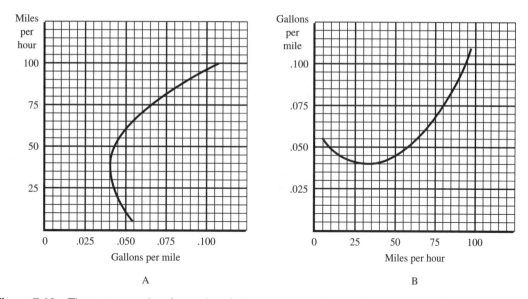

Figure 7.16   These two graphs of speed vs. fuel consumption demonstrate the importance of carefully selecting the independent variable to place on the X-axis.

speed in miles/hour, and the curves are identical, disregarding their differing orientations. Obviously, the graph in Figure 7.16B creates the best mental picture. You will also agree that miles/hour is the independent variable. Less obvious is that the independent (X-axis) variable in Figure 7.16B does not yield more than one value on Y for each data point on X. Can you say the same for the X-axis variable in Figure 7.16A?

When the graph is used to examine points on the line between known data points, the process is called **interpolation.** If the graph is a straight-line graph, interpolation is a simple process (Figure 7.17A).

Interpolation of data along a curved line is another matter. The curved line is your best estimate of what the line should look like. Extracting information between known data points on a curved line is an estimate (Figure 7.17B).

Estimating the curve's fit beyond the experimental data points is known as **extrapolation.** Again, a straight-line graph will offer easy and (usually) reliable results (Figure 7.18A). Extending a curved line (exponential function) requires greater skill (Figure 7.18B). A template (such as a French curve) for drawing curves is useful for drawing curves (both smooth and irregular) of best fit and for extending curved lines.

Curved graphs (e.g., exponential and inverse functions) can be straightened by using semilogarithmic (semi-log) or logarithmic (log-log) graph paper (Figures 7.19A and 7.19B).

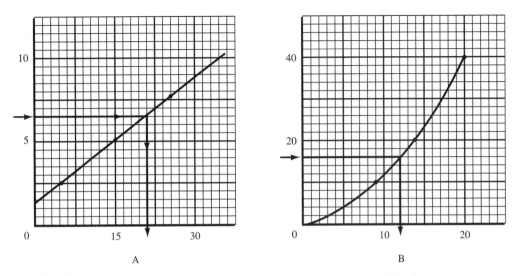

Figure 7.17 Examples of interpolation using (A) the graph of a linear, or straight-line, function and (B) the graph of an exponential, or curved-line, function.

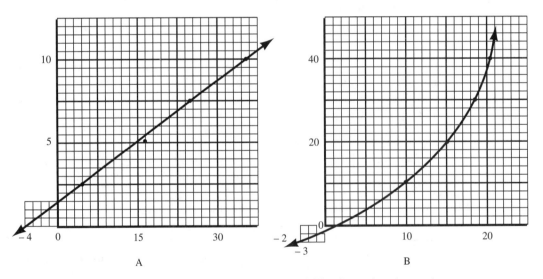

Figure 7.18 Examples of extrapolation using the graphs of (A) a linear function and (B) an exponential function.

You may also be asked to prepare or interpret graphs describing more than one relationship for an independent variable. In Figure 7.20, a motor's shaft horsepower under load is examined. Note that the data points for each line are plotted with a different symbol—circles, squares, and triangles in this case. Use of colors for the different curves is appealing, but when the report is copied the different colors may not reproduce.

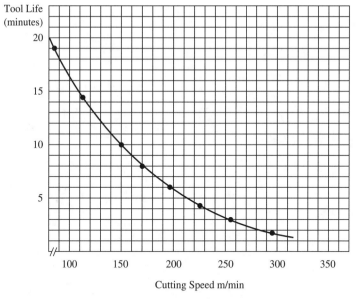

A

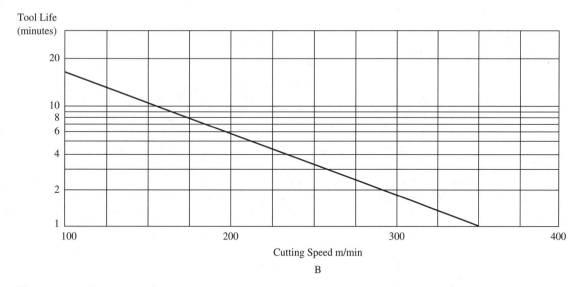

B

Figure 7.19 Tool life depends on cutting speed. Nonlinear graph in (A). Semi-logarithmic paper yields straight-line graph in (B).

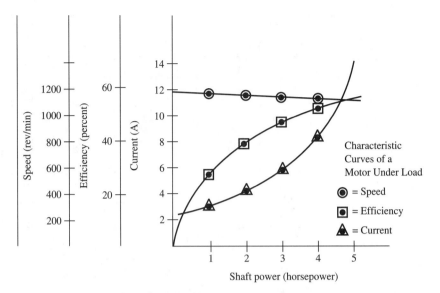

Figure 7.20 An example of a line graph depicting multiple relationships to the one independent variable—horsepower.

7.5 Reporting

When considering reporting we usually think only of written reports. Interestingly enough, oral reports are more common (Figure 7.21). The *oral report,* formal or informal, is important and demands numerous skills—organizational skills as well as speaking and listening skills.

The most frequent reporting we do is informal reporting: e.g., a chat with the boss in the hallway. Anytime we are talking about the progress of our work, we need to follow certain conventions. The manager may seem informal or even flip when inquiring as to how the research or the newly implemented manufacturing process is going, but will be looking for certain insights into the exact nature of what is going on. This means you should, as much as possible, be prepared and factual, avoid going further than what you know, and never exaggerate or embellish. Your career will be enhanced by what you say in informal situations as much as in formal situations.

Ironically, you may best orally report by knowing the major parts of a *written report*—i.e., objective, theory, procedure, equipment usage/problems, data collection and interpretation, and conclusion. By keeping these topical elements in mind, you may be sure you have communicated fully and succinctly what a manager wishes to know. So, by discussing the formal written report we will also be heightening our awareness of what is needed in oral communications.

Figure 7.21   Oral reporting is an important communication tool in business and industry.
(Courtesy Central Ohio Technical College)

Outlining the Written Report

A well-written laboratory report contains a report of all activities leading to a logical and
believable conclusion. The report will contain some or all of the following sections:

1. the *objective,* or purpose;
2. a list of the *equipment* used;
3. the *theory,* or principle, examined;
4. the *procedure* used to successfully complete the experiment;
5. a presentation of the *data* collected;
6. sample *calculations* used to arrive at a conclusion;
7. presentation of the *results*—usually a graph; and
8. *conclusions* and recommendations.

It may help you to memorize this list if you remember the sentence "**O**ur **e**ngineering
technician **p**ersonnel **d**on't **c**rash **r**ed **c**ars." The first letter in each word is in the same
order as the first letters of the topics italicized in the list.

This list is not exhaustive. Missing, for instance, is the cover page, abstract, and table of contents, nice to have but often not required, especially in your college courses. Also, you will not always need all of the topical areas listed. At times you will merely have a laboratory report form to complete. But even these briefer, less formal, methods of reporting will be enhanced when you know what is included in the formal report.

Objective Section

A well-respected speaker once offered the following advice to aspiring presenters: "Tell them what you are going to say. Say it. Tell them what you said." The objective statement in a report is a succinct way to "tell them what you are going to say." The objective statement informs the reader why you are doing the experiment, or explains the principle to be examined.

Here are some examples of well-written objectives:

1. The purpose of this report is to determine the change in velocity (acceleration) of a free-falling object and determine the local acceleration of gravity (g).
2. The following report is designed to explain how a push-pull transistor amplifier is constructed and to describe the input vs. output of the amplifier.
3. This report will examine the tensile strength of SAE 5130 steel. Also, the breaking strength of the 5130 steel samples will be examined at different temperatures.

Many times the experiment assigned by the instructor will state the objective, but you may be asked to state it in your own words. In any case, be brief and do not attempt to describe the entire experiment in the objective statement.

Equipment Section

The list of equipment informs the reader of the characteristics of the equipment used. Such questions may be answered as "How accurate is the equipment? What sensitivity are the scales? How precise are the measurements?" Sometimes the serial numbers are reported. Then, if the experiment must be repeated or continued at a later time the measurements can be made with the same equipment. Also, if the experimental results are questionable, the equipment used may be checked for needed calibration or maintenance.

An example of an equipment reference is Fluke digital multimeter, model #77, serial #41220694, with an accuracy on the 32 VDC scale of $\pm(0.3 + 1)$ [0.3 percent of full scale $\pm$ 1 digit] and a precision of 0.01 V.

Theory Section

The theory section consists of a discussion—sometimes brief, at other times, detailed—of the basic principles to be examined. If a theory section is included with the instructor's handout or laboratory assignment, there usually is no need to rewrite it in the report. However, you should read this section thoroughly before performing the experiment. It will give you invaluable direction and define the variables you will be measuring.

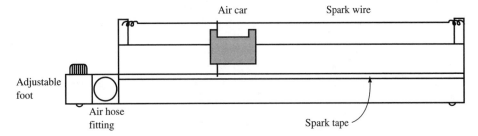

Figure 7.22   An equipment diagram for a frictionless air track.

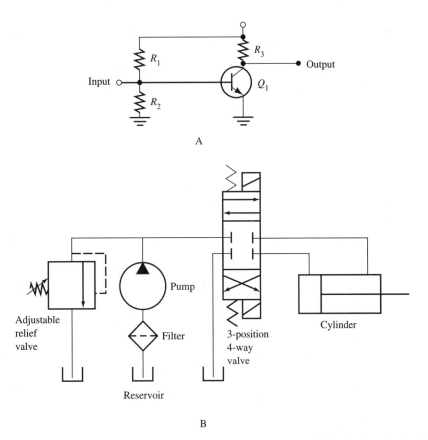

Figure 7.23   Schematic diagrams of (A) a transistor amplifier and (B) a hydraulic circuit.

Procedure Section

In the procedure section you will describe how the experiment will be accomplished. A diagram of the equipment is usually provided, or you may have to draw the diagram (Figure 7.22). A schematic diagram, mechanical or electrical, may be required. See Figures 7.23A and 7.23B for examples. Note the labeling of each part on the schematic.

Step	System pressure		Cylinder force
2	Maximum	PSI	lb
3	Minimum	PSI	lb
Step	System Flow		Cylinder Speed
4	Maximum	gpm	fps
5	Minimum	gpm	fps

A

R_T Required (ohms)	Combination of Parallel Resistors			Measured Value R_T (ohms)
	R_1 (ohms)	R_2 (ohms)	R_3 (ohms)	
500				
1000				
1500				
2000				

B

Figure 7.24 Well-designed forms are used to collect data. (A) Data collection form for hydraulic experiment. (B) Form used to collect resistance measurements.

Neatness and symmetry are also important. For very neat drawings in more formal reports, templates and other drafting aids should be used.

Often, a step-by-step procedure statement is required. If one is offered, it often does not need to be recorded in the report. But to develop a mental picture of the overall process, you should always read the entire procedure before beginning the laboratory. You will have a better chance of performing each step correctly if you visualize the whole process first.

If any portion of the procedure is changed or modified during the performance phase, note each change and describe the modified procedure. When procedure with changes is documented, the experiment can be replicated if necessary.

Data Section

Collected data *must be recorded in a neat and orderly fashion.* This point cannot be stressed too much, and sound data collection is facilitated by a well-designed form. Many times data have been lost or confused because a good data collection form was not used. The form in Figure 7.24A is an example of a data collection form keyed to the procedural steps of the experiment. Figure 7.24B is keyed to the experiment's dependent variable. The appropriate combination of resistors R_1, R_2, and R_3 will result in the required R_T in ohms.

You may use pencil to record data in the college laboratory. Many companies require data to be taken in pen and ink so data cannot be manipulated after the experiment.

When collecting data, be careful in using significant figures and rounding results. Show a good understanding of the units involved. It is a good idea to be consistent with the units used throughout the report. For instance, if millimeters are used at one point, thousandths of an inch should not be used at another point in the same report.

Also, realize that *you alone are responsible for the accuracy and completeness of the data,* even though others may be involved in the experiment (see Example 3.1). It is a good idea to check your data against your partner's data. It is frustrating to find data missing after the laboratory is completed and you are writing the report.

Calculations Section

You should present sample calculations used to determine values of dependent variables and test parameters (e.g., percent error calculations) in the experiment. Sample calculations should follow a logical (chronological) sequence. Do not write repeated calculations. Show each type of calculation only once.

For example, determine the relationship between the horizontal velocity (V)—independent variable—of a projectile at a fixed height (h) above a flat surface, and the range (R)—dependent variable—of the projectile.

$$R = V \sqrt{2h/g}$$

where R = range, V = initial velocity (use 25 ft/s), h = height of projectile (use 65 ft), and g = local acceleration of gravity (32.2 ft/s^2). Substituting the given quantities:

$$R = 25 \text{ ft/s} \times \sqrt{\frac{2 \times (65 \text{ ft})}{32 \text{ ft/s}^2}} = 50 \text{ ft}$$

Results Section

The presentation of the results clearly shows the relationship of two or more variables. It often takes the form of a graph (see Section 7.4). It always takes the form of a drawing, picture, or other visualization, so the reader can see the relationships involved. Figure 7.25 shows how the vector sum of two forces in newtons is determined by the vector-triangle method.

Conclusion Section

The conclusion is the most important part of the report to the reader. Managers, who frequently read reports, always turn to the conclusion first. If nothing of interest is

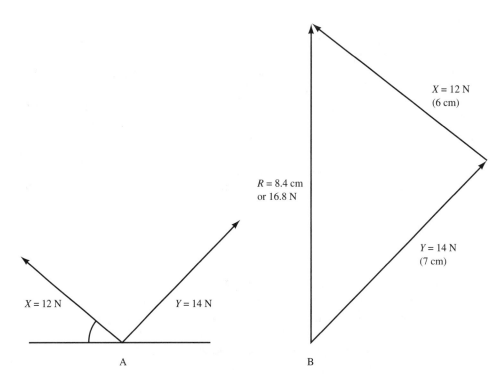

Figure 7.25 The vector-triangle method is used to add two vectors, $X + Y$. Vector X is moved to the head of vector Y, and the resultant (R) is drawn to connect the two. Scaling R and converting to newtons (N) of force yields a vector sum of 16.8 N.

apparent, they simply will not read the report and may think the author is not appropriately addressing the "real concerns" of the enterprise.

The conclusion is where you "tell them what you said." However, do not read off a laundry list of mistakes and attempt to describe the arduous journey taken to collect the data. And, while a student, do *not* begin with "I learned a lot." It is better to write the report as if you were reporting to a manager in industry. The reader wishes to know such things as

- What was the objective of the experiment?
- How well did you achieve the objective?
- How accurate are the data?
- What errors did you encounter during the data collection?
- What was the percent error (Table 7.1)?
- How could the experiment be improved?
- How can the results be applied?

All of these questions need not be answered in every conclusion. Use them as a guide to help you write a thorough conclusion. Inexperienced report writers find this section the most difficult to complete. Also, it is important to not say anything that cannot be backed up by the content of previously written sections of the report.

Recommendations for improvement are often located in a separate section of the report. Many instructors feel this is an extremely important part of the report. Your responses show how much you have thought about the experiment and the errors involved and—more importantly—what you could do to minimize errors or eliminate them altogether. The recommendations section gives you a chance to exercise your practical, creative talents.

Preparing and Submitting the Report

It is a good idea to learn keyboarding and word processing (Section 8.3) early in your career as a technical student. Once you can write on the computer and retain your reports on disk so changes can be readily made, you will never wish to report in any other way. Neatness will be improved and you will be much more productive. Spelling and grammar are also important to good report writing. Most word processing software has a very useful spell-check feature.

Finally, complete your reports on time. Timeliness is difficult for many; they continually want to polish the report and never submit it on time. But deadlines in the workplace are crucial; your instructor expects you to develop good work habits and will demand prompt submission of reports. Penalties will be imposed on those who do not comply.

The Report Checklist

Use the following report checklist in editing your reports:

- The report is attractive. (Neatness does count and papers with ragged edges, torn from a notebook, should never be submitted.) The cover page is well laid out, including such items as your name, the date, the title of the laboratory, other team members' names, and the course section identifier.
- The abstract or objective lets the reader know immediately what the report entails.
- The body provides all essential information, such as the list of equipment with equipment diagrams; a clear, concise procedure supported by theoretical considerations; a good data presentation well supported by attractive and appropriate graphs; and example calculations with literal equations that are representative and in order of use.
- A conclusion that reveals the results obtained, errors encountered, and recommendations for improvement.
- A report that is concise. Conciseness, ironically, takes extra time to achieve, but is always desirable in technical reports.
- A well-proofed report, free from typographical errors (typos), misspelled words, and poor grammar. This last check will become easier after you have successfully completed the several communication courses required of all graduates. Be concise and avoid excess verbiage—when proofing, take out all that does not directly apply to the problem under investigation.

Figure 7.26 The presenter uses a computer to generate his visual aids during a process-improvement team meeting. (By permission of ABB Process Automation, Inc.)

Oral Reporting

You will have to give presentations to many different audiences during your career as a technologist. The more successful you are, the more presentations will be demanded of you. For example, the results of problem-solving team meetings must be presented (Figure 7.26). Managers will ask that you present your ideas on how to improve the company's processes. You might act as a trainer on a new process involving complicated equipment. As a supervisor or manager you will be expected to present on such topics as safety, quality, and productivity in the workplace. Informal oral reporting was discussed at the beginning of this section. But the technician must also know how to prepare and deliver formal oral reports.

The previous material in this section is important to the preparation of the oral report. You will need to know, for instance, the outline of the written report and should be familiar with the elements of the "Report Checklist." Most importantly, be aware that you must be as relaxed as possible during your presentation. *The best way to be relaxed is to be prepared.*

Begin your preparation by knowing your audience. You do not want to use complex technical terms with skilled workers who do not have your education. (On the other hand, you do not want to act as though you have the skills those workers possess or that you are somehow "better than they.") You will wish to use your technical language skills and avoid too much detail when presenting to managers who probably have access to much of the information you will share.

Do not completely write out your presentation, but outline it. Be sure to place important facts and data where you can easily access them as needed. Practicing in front of a mirror or with a friend or spouse can help you use the outline most effectively.

Use visual aids to bring focus to the content of your presentation. In manufacturing industries a flowchart of the process flow as shown in Figure 7.26 might be appropriate. In both service and manufacturing industries, data should be presented in visual form. The data may be placed in "consumable form" by using displays such as line graphs, histograms, and pie charts. Many of these data displays can be easily constructed by using available computer software.

A key consideration is knowing how much time you have for your presentation. Most inexperienced presenters worry about not having enough material, when usually the problem is having too much! You want to think out and zero in on the most important objectives, and *never* go over the time scheduled for your presentation. Also, be sure to allow time for questions. Answer all questions as briefly and honestly as possible.

When you are well prepared you will be more confident in your presentation. More confidence will help you to speak slowly and clearly. Your nonverbal gestures will be smooth, natural, and well coordinated with your material. When waiting to give your presentation, it is helpful to practice deep breathing techniques and to alternately tense and relax the muscles in your arms and legs. And of course, continually remind yourself that *you will do a great job.* During the speech, never draw attention to any negative material, such as shortcomings in yourself or others.

Oral Report Checklist

Use the following checklist in preparing to present an oral report:

- Find out, or somehow determine, how long the presentation should be, who the audience is, what the facilities are, and what visual aids will be available.
- Prepare a thorough, easy-to-read outline, keeping in mind the time that is available and leaving suitable time for questions. Use earlier information from the *written report,* in preparing the outline.
- Research facts and collect data to support the presentation.
- Prepare the visual aids, especially data presentations.
- Practice the outline, with the visual aids.
- Dress appropriately for the day of the presentation and use relaxation methods prior to the actual presentation.

Good oral and written reporting is never easy. It requires thoroughness, good attention to detail, great organizational ability, strong analysis skills, and polished communication skills. These characteristics or qualities seldom occur naturally in any of us, but they can be developed. Good reporting may lead to promotion and, perhaps more importantly, a greater contribution to society.

Problems

1. Be able to list from memory the four major attributes of a "good" laboratory technician (see beginning of chapter). Using a 1–5 scale, rate yourself on each attribute.

Section 7.1

2. You are performing an experiment with two others. You are reading the meter scale, Joan is varying the voltage, and Tom is taking the data. Who is responsible for recording the data correctly?

Section 7.2

3. Classify the following errors as gross (G), systematic (S), or random (R). In some cases more than one classification applies.

Type of Error	*Classification (G, S, or R)*
a. meter incorrectly zeroed	G, S
b. not following approved procedure	
c. oscilloscope attenuator controls not detented (locked into) in the cal mode	
d. micrometer caliper not zeroed	
e. sloppiness in an instrument's adjustment	
f. drafts	
g. friction	
h. friction in inclined-plane experiment where lubrication and other variables are constant	
i. reading a meter scale from the side (always the same side)	
j. dirt in a balance scale pivot	
k. failure to level a surface plate	

Section 7.3

4. Determine the number of significant figures in each of the following data.

a. 154.65	**b.** 300.85	**c.** 940.01	**d.** 80 000.2
e. 80 000	**f.** 0.0017	**g.** 8.000 65	**h.** 4000.
i. 5500	**j.** 5500.	**k.** 1800.0	**l.** 4000.0
m. 0.0204	**n.** 0.2100	**o.** 0.009 140 0	

5. Determine the number of significant figures in each of the following data.

a. 2700	**b.** 4400	**c.** 490.06	**d.** 8700.0
e. 60 000	**f.** 0.0071	**g.** 0.7800	**h.** 7000.
i. 200.58	**j.** 658.93	**k.** 60 000.7	**l.** 0.0905
m. 3000.0	**n.** 3.000 56	**o.** 0.002 790 0	

6. Convert the data in Problem 4 to scientific notation, retaining the correct number of significant figures.

7. Convert the data in Problem 5 to scientific notation, retaining the correct number of significant figures.

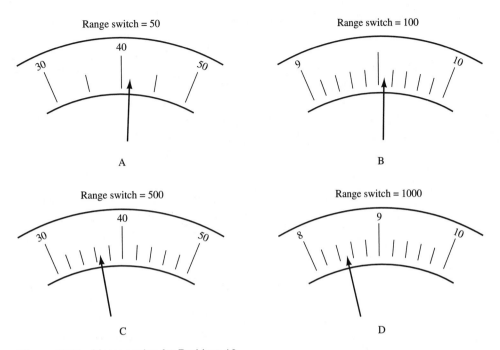

Figure 7.27   Meter scales for Problem 10.

8. You were informed by the instructor that you would receive a C in the course if your numerical course average is 72.5 ± 9.5. What are the maximum and minimum averages that will result in a grade of C?

9. The accuracy of an analog meter is ±5 percent of full scale. What are the maximum and minimum readings for a scale reading of 84.5 and a full scale of 100?

10. Many scales depend on a range switch to determine full scale values (e.g., a scale that ends in 10 with the range switch on 1000 will read 1000 on full scale). Record the readings of the meter scales in Figure 7.27 to the correct number of significant figures, and specify the resolution of each reading as ± the recorded reading. Example: A, 40 ± 2.5. (The 2.5 represents one-half of the smallest division—in this case, 5.)

11. For the digital multimeter readings with the associated accuracies shown below, write the maximum and minimum values of the measurement.
    a. 1000 ± (0.5 + 2)   *Example:* 993.0 to 1007
    b. 500 ± (0.2 + 1)
    c. 40.0 ± (2 + 1)
    d. 32.02 (mA) ± (0.3 + 2)

12. Round the following to three significant figures (use rounding rule 3 in Section 7.3).
    a. 326 413   b. 804.007
    c. 13.455   d. 0.006 075 7

13. Calculate the following to the correct number of significant figures. Use rounding if necessary.
   a. 199×0.3
   b. $95 \times 196 \times 53$
   c. 0.32×0.3
   d. $0.000\ 302 \times 30.7$
   e. $\dfrac{1000}{10}$
   f. $\dfrac{0.034\ 02}{50}$

14. Calculate the following to the correct number of significant figures. Use rounding (rule 1) if necessary.
   a. $530. \times 0.5511$
   b. $8006 \times 68\ 000$
   c. $\dfrac{0.048 \times 880}{60020}$
   d. $\dfrac{13 \times 10^4}{3 \times 10^{-3}}$

15. Calculate the following to the correct number of significant figures. Use rounding if necessary.

a. 3060	**b.** 4000.0	**c.** 0.430	**d.** 6001
230	$-\ 0.3226$	$+\ 3.6$	588
$+40$			$+773$

*16. Calculate the following to the correct number of significant figures. Use rounding if necessary.
   a. $8 \times 10^3 + 1.003$
   b. $6.0329 \times 10^3 - 1.01 \times 10^2$

17. You carefully measure the tensile strength (stretching strength) of five steel specimens in megapascals (MPa). (A pascal is a pressure in newtons per square meter $[N/m^2]$; 6.9 MPa = 1000 PSI.) The measurements are 34.45, 34.40, 34.39, 34.42, and 34.50. Compute the average tensile strength, and express your answer in the appropriate number of significant figures.

18. In Problem 17, compute the percent error of each of the data points from the average (see Table 7.1—use absolute value in numerators).

Section 7.4

19. The graphs in Figure 7.28 are incorrectly drawn. Use the graphing checklist in Section 7.4 to identify the errors. Two errors exist in Figures 7.28C and 7.28D.

20. The following data were recorded during an experiment. Graph the data using the basic principles outlined in the text.

Revolutions per Minute (RPM) of Pump	Gallons per Minute (GPM) of Pump Output
0	0
100	0.5
220	1.4
310	2.6
400	4.2
498	6.0

21. From the graph in Problem 20, find the pump output at 250 RPM by using interpolation.

*Challenging problem.

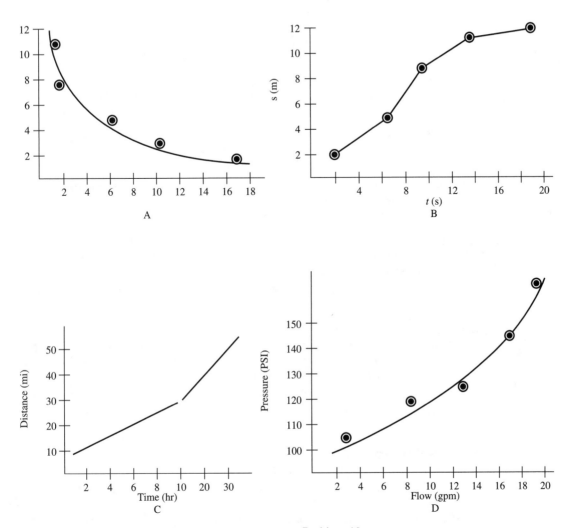

Figure 7.28 These incorrectly drawn graphs support Problem 19.

22. A 10 kΩ potentiometer (variable resistor) is placed across a 10 V battery, varied in 1000 Ω steps from 2000 to 5000 Ω, and measurements taken also at 7000 and 10 000 Ω. Graph the resultant current changes as tabulated below.

R (ohms)	I (mA)
2000	5.68
3000	3.42
4000	2.57
5000	2.57
7000	1.46
10 000	0.996

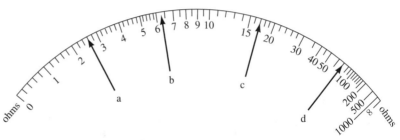

Use scale multiplier of $R \times 1$

Figure 7.29 Reading a nonlinear ohmmeter scale is more difficult than reading linear scales. Figure for Problem 24.

23. From the graph in Problem 22, find the current flow at 11 000 Ω by using extrapolation (an estimate will do fine).

24. Interpret the ohmmeter scale in Figure 7.29 and record readings for a, b, c, and d.

Section 7.5

25. Perform a laboratory exercise, given a list of required equipment, the theory or principle to be examined, and a procedure to be followed. Write a laboratory report and include the particular sections assigned by the instructor.

26. Prepare and deliver a five-minute oral report concerning a technical topic to be approved by your instructor.

Selected Readings

Houp, Kenneth W., and Thomas E. Pearsall. *Reporting Technical Information.* 7th ed. New York: Macmillan, 1992.

Weisman, Herman M. *Basic Technical Writing.* 6th ed. Upper Saddle River, New Jersey: Prentice Hall, 1992.

8

The Personal Computer

The personal computer (PC) has profoundly impacted our lives. The PC has decreased in price to the point where almost everyone can have a home computer, and many people do.

Decreasing cost, increased power, and improved connectivity have resulted in constantly increasing use of the PC in business and industry. Today's competitive industries realize the computer age is not coming, but is here now.

To compete in the computer age, technicians and technologists must

- use appropriate computer terminology when communicating with others;
- understand specific application software: for instance, word processing, spreadsheet, data base management, and drafting software (CAD);
- use a computer keyboard to input information to a PC;
- know how to save information to a disk;
- be able to retrieve previously saved information from a disk; and
- be familiar with BASIC programming.

You can meet these objectives by using this chapter and the next, a PC with a user's manual, and the guidance of your technical instructor. Without access to a PC, you can only meet the first objective.

8.1 Mainframe Computers Lead to the PC

Before proceeding further, reflect on the evolution of the PC. Table 8.1 lists the major categories of computers in use today.

Table 8.1   The Family of Computers

Mainframe computer	The earliest business and industrial computers were built in large mainframe configurations with electronic chassis, or "equipment racks." PCs, now, are as powerful as early mainframes that occupied large rooms and cost millions. Mainframe computers are used today in large-scale data-processing departments (Figure 8.1) and to supervise the actions of smaller computers in manufacturing.
Supercomputers	The supercomputer has replaced the mainframe in solving complex science and technology problems. The supercomputers in Figure 8.2 show the continual downsizing of computers with improving technology. As you move back in the photograph, you find the smallest supercomputer, the Cray-2 (top right-hand corner), to be the most powerful of all. All supercomputers use multiprocessing to execute instructions simultaneously. The Cray-2, with four processors, makes 250 million computations each second (see also Figure 10.2).
Minicomputer	The minicomputer is the forerunner of the PC. It is generally faster, larger, more powerful, and costs a great deal more than the PC. Connecting several PCs to one "mini" (networking) has led to computer-integrated manufacturing (CIM).
Personal Computer	A single *microprocessor* chip acting as the central processing unit (CPU) distinguishes the PC from its predecessors. The PC is fully programmable, has memory, and provides inputs and outputs just as the minicomputers and mainframes do. The reason for the PC's popularity and greatly increased use is its lower cost—the average family can easily afford a PC. Figure 8.3 depicts a popular PC.

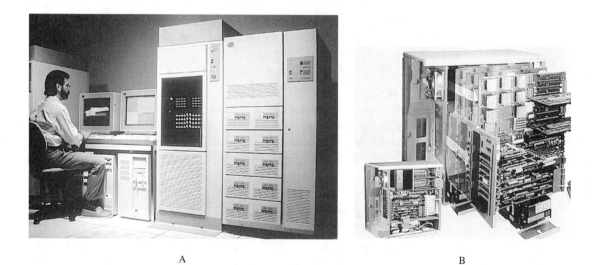

A   B

Figure 8.1   (A) One of the industry's most advanced mainframe computer systems. (B) An exploded view of the mainframe. (Courtesy of International Business Machines Corporation. Unauthorized use not permitted.)

Figure 8.2 One of the world's most powerful arrays of supercomputers. From left (bottom), Cray X-MP/22; a Cray-1S and Cray-1; and at the right (top), the small Cray-2 is a powerful four-processor computer. (Courtesy Lawrence Livermore National Laboratory)

All the computers listed in Table 8.1 can be defined functionally as follows:

1. A computer accepts information (input).
2. It manipulates the data by following a predetermined set of instructions in memory (processing).
3. It communicates the results in consumable form as information (output).

Figure 8.3 A popular PC. (Courtesy of International Business Machines Corporation. Unauthorized use not permitted.)

Figures 8.4A and 8.4B show, in block diagram form, a human's interaction and a computer's interaction with the environment. It is easy to see the relationship between our brains, the processor for our bodies, and the computer's processor. The five senses are equivalent to the inputs to the computer processor. Our muscles attached to bones allow the brain to create outputs for our interaction with the environment.

Your first encounter with a computer processor, or central processing unit (CPU), was probably an arcade game. A coin-operated video game will be used to further illustrate the parts of the CPU. When you place a coin into the slot, the computer accepts this as an input. After the input is accepted by the CPU, processing takes place, and (if you paid the correct amount) the game is sequenced to start. A predetermined set of instructions (application software) sets up the gameboard format on the video screen. The video screen and the sounds of alien aircraft firing laser cannons are usable outputs. You then input, via joystick, trackball, or fire-button. The CPU accepts these inputs and then outputs by creating moving images and sounds and displaying the score. When the information to the processor meets the criteria for the end of the game as established by software stored in memory, the processor will output "Game over."

EXAMPLE 8.1   Draw a flowchart of the arcade game's sequence of events.

Solution   We will use only four symbols in the flowchart. (Programmers use up to eight symbols or more.) A parallelogram indicates inputs and outputs, a rectangle indicates a processing function, the diamond represents a decision, and an open-ended

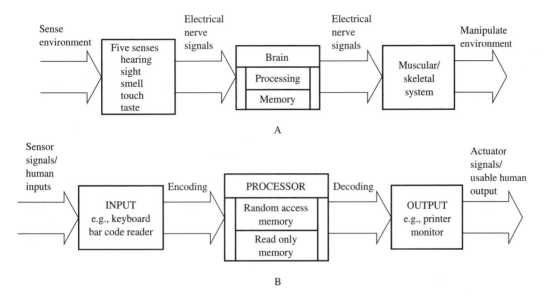

Figure 8.4 The block diagrams illustrate (A) a human's interaction with the world and the processing of information and (B) the computer system's interaction with the world.

rectangle indicates information stored in the computer. Of course, connecting arrows show the sequence of events. Figure 8.5 is the resulting flowchart for the arcade game. Can you find missing steps on the flowchart that could be added?

The PC is made possible by the development of the microprocessor. The **microprocessor** is a single integrated circuit (IC) often called a microprocessor *chip*. The chip is the CPU for the PC (Figure 8.6). We find the chip, or CPU, in telephones, automobiles, calculators, and many home appliances. The microprocessor chip, then, is used not only in a PC, but is used to control instruments as well.

As an example, one type of 35 mm camera uses a microprocessor for determining exposure setting (diaphragm adjustment). The microprocessor is only 3 mm^2 in size.

Microprocessors in automobiles control spark plug timing to enhance fuel efficiency and minimize exhaust pollution. The chip samples many engine conditions before making a decision as to the correct spark timing. The monitored conditions (inputs) include RPM, temperature, and manifold absolute pressure. Tables stored in memory (often called *look-up tables*) are compared with incoming data to determine the advance or retardation of the spark.

EXAMPLE 8.2

Block-diagram the automobile's microprocessor control of spark plug timing.

Solution   See Figure 8.7.

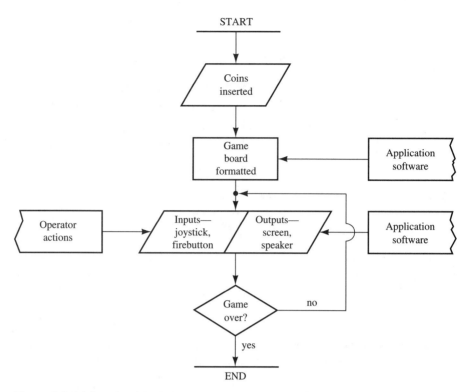

Figure 8.5 A flowchart depicting the arcade game's sequence of events.

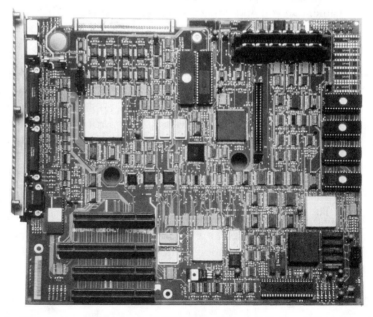

Figure 8.6 The system board for the IBM Personal System/2. The CPU for the board is the dark square chip in the lower right-hand corner. (Courtesy of International Business Machines Corporation. Unauthorized use not permitted.)

A

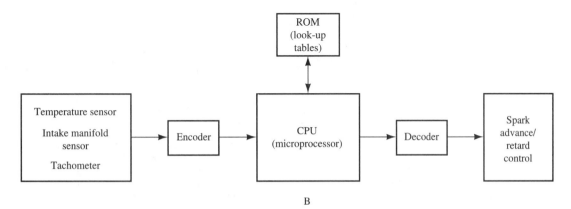

B

Figure 8.7 The block diagram of an automotive computer designed to control spark plug timing. (Courtesy of International Business Machines Corporation. Unauthorized use not permitted.)

Now that we have seen some applications for the microprocessor, let us examine our innate human concerns about being a part of the information age. We all share such concerns and many of us feel nervous about facing the tremendous challenge of the world of the computer.

The information age brought on by the computer has ballooned our information base. Terms never encountered a few years ago are suddenly necessary vocabulary words. We must understand these new expressions if we are to communicate with others in business and industry. Some of the computer expressions you must be familiar with and use properly are *hardware, software, firmware, data base, floppy disk,* and *hard disk.* New and relatively large dictionaries are devoted solely to the computer.

With characteristic brusqueness, this information world also uses an excess of acronyms and abbreviations such as IC, PC, DOS (*d*isk *o*perating *s*ystem), CPU (remember this one?—see Figure 8.8), ROM, RAM, BASIC, and so on. Formidable as the new language is, *to cope in today's computer world, we must be fluent with its terminology.*

Refer to the following brief glossary often as you read on. Because the terms used are vital to the technician, you may be asked to define all glossary words introduced in this chapter.

Glossary 8.1

CENTRAL PROCESSING UNIT (CPU): The central processing unit contains the control unit and the arithmetic logic unit (Figure 8.8).

DISK: Information on RAM may be stored permanently on magnetic disk. The permanent record on disk may be "read back" into RAM at any time.

DISK OPERATING SYSTEM (DOS): The disk operating system formats disks and manages disk operation. MS-DOS is a trademark of the Microsoft Corporation, and PC-DOS refers to the disk operating system for a large variety of look-alikes or

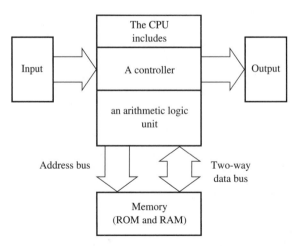

Figure 8.8 The functional block diagram of a CPU. A "bus" refers to two or more conductors carrying information to or from the CPU.

clones (personal computers based on the original IBM operating system). The MS-DOS has a *command orientation,* which is not as easy to use as systems using a *menu orientation,* such as the Apple Macintosh and Microsoft Windows operating systems.

HARDWARE: The computer housing and all parts it contains are called hardware, e.g., the CPU and memory devices. Hardware, as distinguished from software, is anything you can see or feel. Hardware also includes peripherals, such as disk drives, screens, and printers.

MICROCOMPUTER: A computer that uses only one microprocessor or CPU. *The microcomputer is the same as a PC.*

MICROPROCESSOR: A single-chip CPU. The terms *microprocessor* and *microcomputer* should not be confused.

PERIPHERALS: Peripherals include all equipment and devices external to the computer, but necessary to interface to the real world and the computer operator (e.g., input devices—keyboard, barcode reader, joystick, and output devices—printer, monitor, speaker).

PERSONAL COMPUTER (PC): A popular term, personal computer is used to describe a general, off-the-shelf microcomputer. Also used to describe IBM-based computers, as opposed to Macintosh-based machines.

RANDOM ACCESS MEMORY (RAM): Random access memory is sometimes referred to as "scratchpad memory" because the processor can write information into RAM locations for later use, or it can "scratch" the information and write new data into the same location. Disks are used to write application software (see below) into RAM.

READ ONLY MEMORY (ROM): Read only memory is installed by the manufacturer and cannot be erased or written over. Some system software (see below) is located on ROM, so it cannot be accessed by the operator or programmer. When the computer is first "turned on," the CPU uses ROM to access its initial instructions.

SOFTWARE: Software is information and cannot be seen or felt. Nevertheless, it exists as an important entity. Software makes the world of the digital computer possible and is what users directly interact with. Software is divided into two major categories:

- **system software** instructs the processor how to interact with the inputs (e.g., keyboard) and outputs (e.g., screen and printer), and
- **applications software** is brought into the computer via the disk drive for specific applications (e.g., word processing, spreadsheet, data base management, and drafting and design [CAD]).

Coupling the vast amount of new terminology with the rapid, continuous expansion of technology, we find a world quite complex and difficult to compete in. It seems overwhelming at first, but after a few courses at your two-year or technical college you will begin to see how easy most of the new terms are to understand, once they are coupled with applications. This is where the technical college is most responsive to business and industrial needs—keeping up with the state of the art in information processing and computer manufacturing. Many four-year college graduates enroll in technical colleges to take advantage of up-to-date computer courses with state-of-the-art equipment.

Problem-solving skills were discussed in Chapter 3. To handle this complex and rapidly changing world, you can decide to focus on certain applications and use good decision-making skills to select those aspects of the computer to concentrate on to get what you want from the technology. You also must constantly remind yourself that everyone else is struggling to keep up too. So, when you feel overwhelmed, remember that everyone feels the same way. Also remember, the computer is increasing productivity and improving our standard of living—we cannot live without it.

You must learn all you can about the digital revolution during your college education and beyond. You will have no problem finding current information in your area of interest. If anything, the problem is too much information. All new technologies enjoy an explosion of literature at first. After more standardization takes place, a few strong reference sources will be left.

For now, enjoy the fast pace and revolutionary change. Always be sensitive to those around you who do not understand the computer and computer-related terms as well as you. Be open to their ideas on how the technology may be made clearer—"friendlier"—to all. As a technician you can create change, improve understanding, and make other contributions to this volatile field.

8.2 The Personal Computer System

The personal computer is a complex system. You will work with a lot of complex systems in technology. No matter how complex the system, it is easier to understand if you are familiar with its basic building blocks. The best troubleshooters isolate problems in a complex system by block diagramming the system. Figure 8.9 shows three building blocks of a PC. The hardware is no more than a typewriter (the keyboard) with a videoscreen output. Whatever is typed on the keyboard will appear on the screen. The ROM is permanently stored memory on a silicon chip and remains even when the computer is turned off. One of the important things ROM does is to tell the microprocessor what to do when the power comes on.

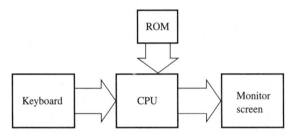

Figure 8.9 A computer typewriter with videoscreen output. (This block/diagram format will orient you to the basic parts of the computer. It will be expanded.)

Actually, there is a lot happening even when the computer is only printing on a screen what is typed on the keyboard. For instance, typing an X on the keyboard sends an electronic signal to the microprocessor. The CPU works with the system software in ROM to identify the X and convert the signal to a **machine language** consisting of only 1 and 0. This is the only language the CPU understands. Computer languages will be covered in more detail in Chapter 9. The CPU and the ROM (containing system software) then convert the machine language to information the screen understands. After the ROM does its conversion, an electronic signal is sent to the screen as an X. If this sounds like a lot happening, remember that all of this is accomplished in microseconds, at least a hundred thousand times faster than you can think up the next letter to type!

In Figure 8.10, two more blocks are added to the complex system started in Figure 8.9. Most new computer users begin with software someone else has written (programmed) to do a specialized job. This task-oriented software must be distinguished from the system software and is known as **application software.** In a personal computer the application software is installed in RAM by disk drive. Unlike ROM, RAM is transient (lost when the PC is turned off) and can be changed or modified.

Personal computers have two or more disk drives. The two types of disk drives are known as "floppy" and "hard" drives. (CD-ROM and other optical drives are discussed in Section 10.3.) Hard drives, discussed later, have fixed disks located inside

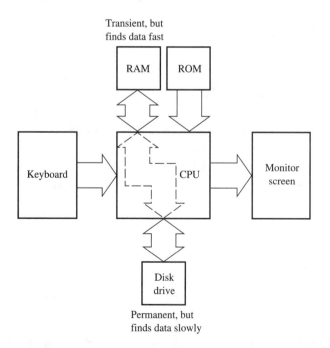

Figure 8.10 The PC's disk drive reads instructions from application software into the random access memory (RAM).

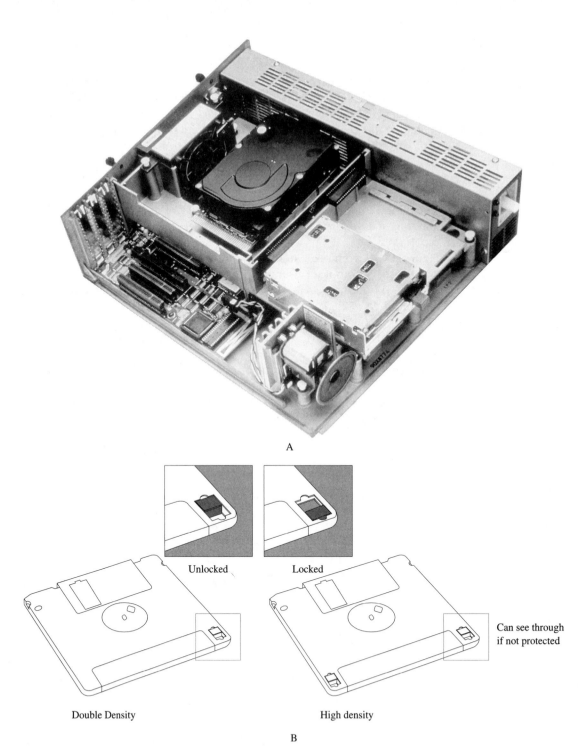

A

Unlocked   Locked

Can see through
if not protected

Double Density   High density

B

Figure 8.11 (A) A floppy disk drive (diskette drive) for the PC in Figure 8.3. The drive is located above the speaker in the lower right-hand corner of the photo. (Courtesy of International Business Machines Corporation. Unauthorized use not permitted.) (B) Double-density and high-density 3.5-inch floppy disks.

the PC and are inaccessible to the user. The hard drive usually contains the application software, such as the word-processing software, described in Section 8.3. Documents that result from using the application software can then be transferred to a floppy disk.

Floppy disk drives (Figure 8.11A) are accessible to the user and the disks can be removed, stored, and used to transfer information from one PC to another. The 3.5-inch disk contains either 720 KB (720,000 bytes) or 1.44 MB (1,440,000 bytes) of information (one byte is 8 bits of information). Excluding the photos and drawings, a single 720 KB disk will hold the information in this entire text! The 720 KB disk is known as a double-density (DD) disk, and the 1.44 MB is known as a high-density (HD) disk. The high-density drives for both types of disks will almost always read the double-density disks, but *the double-density drive heads are too wide to read the tracks on the high-density disks.* The 3.5-inch disk may be "locked" in order to prevent writing to the disk. This is known as write-protection and is illustrated for both DD and HD disks in Figure 8.11B.

One of the first operations you will be required to do is to **format** a blank disk (see Figure 8.12). You may think of formatting a disk as similar to drawing lines on a paper before you write on it. Each line can represent one "sector." In this way you not only write straight lines, easily read by others, but the reader can also quickly identify what

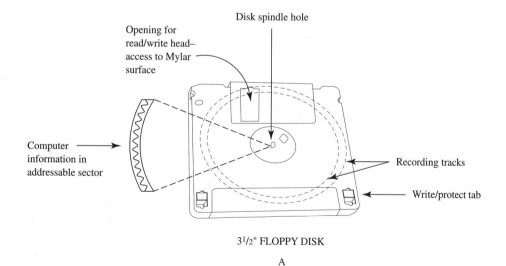

Opening for
read/write head–
access to Mylar
surface

Disk spindle hole

Computer
information in
addressable sector

Recording tracks

Write/protect tab

3¹/₂" FLOPPY DISK

A

Figure 8.12 (A) Formatting or initializing a disk provides address locations the CPU can find. (B) A highly magnified area of a computer disk surface. One square inch contains 22 million bits. (C) 45 tracks on a high-density disk fit between two lines of a fingerprint.

B

C

Figure 8.12 (*Continued*)

Table 8.2 Selected PC-DOS Key Control Functions (The inside back cover has selected DOS commands)

ENTER	Enters the current line into RAM, so the computer can act on it
ESCAPE	Aborts current line without entering it as a DOS command or a file name
CTRL ALT-DEL	A three-key combination. Press and hold the control (CTRL) and ALT keys, then hit the DEL key. This command erases all data from RAM and reloads DOS. (This is known as the "three-finger solution" to all your computer problems.)
CTRL BREAK	Interrupts any DOS command operation, and returns you to the prompt (C>)
SHIFT-PRTSC	Prints out on the printer whatever is on the screen

line contains a certain item. In the case of the PC, the operating system labels each sector of each track with an address, so a search for instructions can begin in a certain sector rather than having to start over as is necessary with a magnetic tape drive.

Of course, different computer operating systems format disks differently. Computers using the same operating system may have drives that format and read only double-density disks, although most computers will format and read both DD and HD disks. Your instructor is a critical resource as you format your personal floppy disk.

The bottom line is this: "Unformatted disks will not run on any operating system." See Table 8.2 for some PC-DOS key control functions.

```
C:\> VER
MS-DOS VERSION 6.22

C:\>format a:/F:1.44
Insert new diskette for drive A:
and press ENTER when ready...

Checking existing disk format.
Saving UNFORMAT information.
Verifying 1.44M
Format complete.

Volume label (11 characters, ENTER for none)? TECHDATA

    1,457,664 bytes total disk space
    1,457,664 bytes available on disk

          512 bytes in each allocation unit.
        2,847 allocation units available on disk.

Volume Serial Number is 3866-10CC

Format another (Y/N)?N

C:\>
```

Figure 8.13   A screen print of the formatting process of a floppy disk in drive A.

EXAMPLE   Format a new disk on a popular operating system and use the screen copy command
8.3   on the keyboard to verify formatting steps.

Solution   In Figure 8.13 the screen shows the entire formatting process from computer turn-on (cold boot) to formatting completion. **Operator inputs are shown in
bold type.** The "boot" is automatically initiated by the operating system and termi-
nates with the prompt **(C:\>)**, indicating the disk operating system (DOS) software
being used is on the disk in the C drive.
    The following steps complete the process:

1. At the DOS prompt **(C:\>),** the command **VER** is entered to verify the ver-
   sion of MS-DOS on the hard drive. The 22nd upgrade of Version 6 is shown.
   Then type **format a: / F 720** or **1.44,** (use 720 for DD 3.5-inch disks, 1.44 for
   HD 3.5-inch disks).

The screen shows

(C:\>) format a: / F 1.44

2. Press **Enter.** The system software asks for an unformatted disk to be placed into drive **A,** for formatting. (Make sure when you type **format** you always follow the command with a space and then **A:.** Otherwise, you may wipe out all of the information on your hard disk in drive **C.**) The unformatted disk is placed in the **A** drive.
3. Press **Enter** again. The operating system checks the disk for any existing data on the disk if it has already been formatted and used. The existing data will be saved in RAM.
4. The system has completed the formatting and the user may add a volume label of up to 11 characters. The operator gives the disk volume a label of **TECHDATA** and presses **Enter.** The user is informed of the disk space available in bytes. The type of disk just formatted is 3.5-inch, HD, and 1.44 MB.
5. Once the disk is formatted you are asked if you wish to format another disk. Some users format several disks at once. To format another disk, type **Y** for yes. To exit, type **N** for no.
6. Watch the drive light to know when the formatting process is complete. *Never try to remove a disk before the drive light is off.*
7. The PC-DOS Key Control Function (Table 8.2) is SHIFT-PRTSC.
8. You may verify that your label is formatted by using the **DIR** command (inside-back cover). In this case the volume label, *TECHDATA,* will be shown.

Note: Most 3 1/2″ floppy disks now come formatted for DOS.

Windows 95 is an operating environment integrated with DOS. Multiple screens are used in many cases. When formatting a disk using Windows 95, the results are the same as with DOS. The difference is that Windows is a graphical user interface (GUI—pronounced gooey). The GUI allows a mouse to click on pictures called icons to achieve the same results as typing text commands. In other words the mouse is the *clicker* input device. Windows uses two clickers on the mouse, the *left click* and the *right click.* (If there are three clickers, ignore the center one.)

The following steps complete the formatting process with Windows 95:

1. To begin, double-left click on the **My Computer** icon (shown in the top-left corner of the screen). The menu box appears with several icons (Figure 8.14A).
2. Right click on the icon labeled **3 1/2 Floppy [A:].** A pull-down menu appears (also Figure 8.14A) listing several options. The Format command is highlighted.
3. Left click on the **Format** command and a Format box will appear (Figure 8.14B). The box shows several format options.

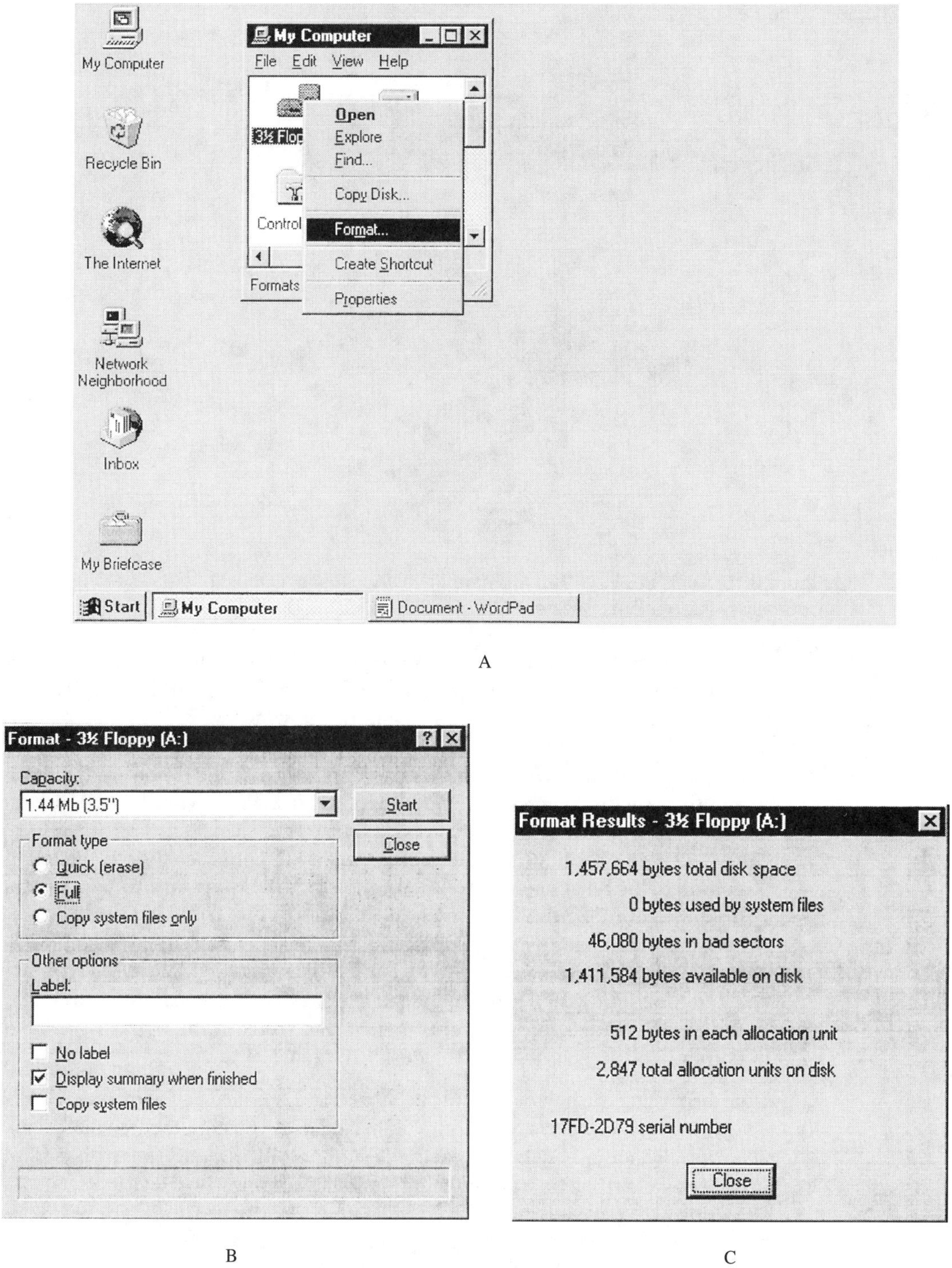

A

B

C

Figure 8.14 Formatting with Windows 95. Computer screens from Windows.

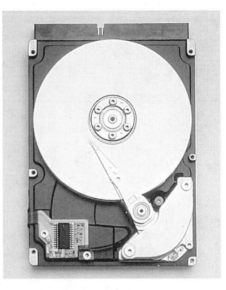

Figure 8.15 This 3.5-inch, 720 MB hard drive should satisfy most of today's PC users. (Courtesy of International Business Machines Corporation. Unauthorized use not permitted.)

Capacity indicates the size of the disk to be formatted. The most common floppy is the default **1.44 MB [3.5″].** For a different disk size, left click on the down arrow located to the right of the window.

Format type asks how the disk is to be formatted. Left click on the **Full** option for all unformatted disks. The default is **Quick** and is only used for previously formatted disks. Remember that formatting a previously used disk removes all information from that disk!

Other options gives the user the ability to **Label** the disk. Left click to complete the box or select **No label.** If selected, **Display summary** displays disk format information as shown in Figure 8.14 C. **Copy system files** creates a "bootable floppy" or a floppy disk that may be used to start a computer from a floppy drive.

4. After completing the Format window, left click on the **Start** button to begin the disk formatting procedure. Left click the **X** box in the upper right-hand corner of all screens to clear the desktop.

After formatting, floppy disks must be stored and handled properly. Keep disks dry and away from extreme temperatures. For instance, do not store them in direct sunlight and especially not in a hot closed car in the summer! Be careful to keep disks from strong magnetic fields that exist around computers, telephones, or television sets.

The floppy disk drive is one type of magnetic disk drive. The 3.5-inch drive is now the standard and few 5.25-inch drives remain. A smaller drive used in laptops is the 2.5-inch drive, with a 1.8-inch drive on the drawing boards.

Today's PC's will have a **hard disk drive** capable of storing much more information than may be stored in a floppy disk drive (Figure 8.15). In this case, at least one floppy disk drive will remain to hold your personal disk. Make sure you always copy your work to a personal disk after creating a file on the hard disk (see Example 8.4), when using your college's computer.

Hard disk drives based on magnetic storage are continuously evolving to increase storage capability. In the future we will see a combination of optical and magnetic storage. **Magneto-optical (M-O)** drives have the ability to store 45 gigabits per square inch. This means that a large textbook could be written on the head of a pin. Standards already exist for a M-O floppy drive to store 650 Megabytes, about the same storage capability as existing CD-ROM drives (Section 10.3).

Disk drives—optical or magnetic, floppy or hard—are known as computer peripherals. A **computer peripheral** is any hardware device connected to the computer's processor. The technologist must be familiar with many types of input and output peripherals. The PC keyboard, for example, is an **input peripheral.** It communicates with the CPU by encoding the inputs into a code the microprocessor understands.

The computer system in Figure 8.3 has a keyboard with a number pad. The **number pad** allows an operator who has learned keyboarding (the touch system) to input numbers while reading data. Most technical colleges offer a keyboarding course and it is highly beneficial. It seems peculiar to see a computer operator or programmer using the "hunt-and-peck" system when working with a machine that processes information in microseconds.

Unlike the keyboard, the disk drive represents a peripheral that can input (encode) and accept outputs (decoded signals). A specific disk drive (most PCs have more than one) needs to be identified by the operating system before it can be written to. The prompt on the screen might look like this: **"A:>"** (read "A prompt"), for drive A:, or **"B:>"** for drive B:, or **"C:>"** for drive C: (usually the hard drive if one is available).

EXAMPLE 8.4

Copy a file from the hard drive to a formatted floppy disk in drive A.

Solution 1   Using DOS

1. If you are already at the C drive, the prompt will read **C:\>** (the backslash [\] indicates you are in a subdirectory of the C drive). If your prompt indicates another drive, type **C:.**
2. Type in

   COPY POLTOREC.BAS A: (NOTE! Place a space between .BAS and A:.)

3. The screen shows

   C:\>POLTOREC.BAS A:

4. Make sure your newly formatted disk is in the A drive and press **Enter.**
5. The computer responds with

   1 FILE(S) COPIED

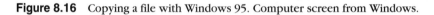

Figure 8.16 Copying a file with Windows 95. Computer screen from Windows.

Solution 2 Using Windows (Refer to Figure 8.16)

1. Insert the disk to be formatted into the floppy disk drive.
2. Open **My Computer** by left clicking on the desktop icon.
3. Open the drive the file to be copied is on, in this case the C drive, by left clicking its icon.

4. Right click the icon of the file to be copied. A pull-down menu appears.
5. Move to **Send to** and another pull-down menu appears.
6. Left click on the drive you wish to copy the file to, in this case the A drive (3 1/2 Floppy [A]).

Output peripherals are monitors (screens) and printers. In industry, the PC's output could drive a number of different transducers, such as a motor or generator.

EXAMPLE 8.5

A laser printer types 10 pages/minute at 25 lines per page and 65 characters/line. Calculate the typing speed in words/minute if a word counts as six characters when typing speed is measured.

Solution Typing speed in words per minute (WPM) is calculated by

$$\text{WPM} = \frac{10 \text{ pages/min} \times 25 \text{ lines/page} \times 65 \text{ characters/line}}{6 \text{ characters/word}}$$

$$= 2708 \text{ words/min}$$

The unit cancellations are left for you.

Other peripherals such as modems (Section 8.4) are used to communicate between your computer and other computers. Many other peripherals could be listed (Figure 8.17). But for now focus on the use of the keyboard, disk drive, screen, and printer.

Figure 8.17   The scanner is an input peripheral. (Courtesy of International Business Machines Corporation. Unauthorized use not permitted.)

Remaining discussion in this chapter assumes that you have access to a PC, the operating system is installed (the system is booted), and you are somewhat familiar with the keyboard. Your instructor has also given you a few of the commands necessary to

- input information through the keyboard,
- make corrections,
- access different disk drives,
- save information to disk, and
- read information from disk to computer (RAM).

8.3 Application Software—A Survey

Basic system components for the personal computer have been described. You now realize the importance of knowing what part of the computer is active—what block you are dealing with (see Figures 8.9 and 8.10). As you begin to use application software, you should ask yourself, "Am I working with the application software or the system software?" When your actions result in an error message, you should immediately attempt to identify what part of the system or application software has been offended. The software will identify what is wrong, but you must interpret the message.

Basic operational skills with the PC and its inputs and outputs have also been established. You should now be able, for instance, to boot a computer, format a disk, save information to disk, and retrieve information from disk. In other words, you can understand and use the operating system for your PC.

With these skills developed, productivity can begin—almost immediately. Application software has been written by career programmers so it can be used easily by anyone. This is what is meant by the term *user friendly.*

When using application software you interact first with the software and then with the PC operating system (Figure 8.18). Because you cannot see or touch software it is hard to think of it as a tangible entity. But information, especially in the computer world, is a very real thing to be reckoned with. The fact is, the most important decision you will make during your career is what software to buy. Computer buyers are urged to always select the application software they need before attempting to identify the hardware and operating system to run the software.

Today's computer systems offer **integrated operating environments** allowing users to transfer data from one program to another. The two most popular environments are Microsoft Windows and Apple Macintosh. Both of these programs work by using windows or icons on the screen to aid the user in operating the software. Most business and industry users have invested heavily in the In tel-based system (Intel manufactures the CPU) that uses Windows. But a significant market share belongs to Macintosh systems, including many schools and colleges.

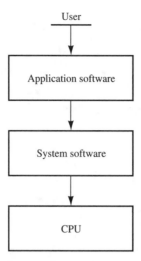

Figure 8.18 When using application software, a user actually interfaces first with the application software, second with the system software, and third with the computer's processor.

In the next three sections, six different types of software that engineering technicians use in their day-to-day tasks will be surveyed. The six representative types of software are

- word processing,
- electronic spreadsheet,
- data base management,
- simulation,
- Internet software, and
- computer-aided design (Section 8.5 addresses CAD hardware and software).

The text will use examples of certain types of application software, but it is not meant to be product-specific. Therefore, you will note an absence of software commands. You must use the **documentation** (printed material such as a user's manual) from the application software to discover appropriate commands.

Word Processing

The typewriter allows us to communicate much faster and more surely than we could with handwriting. With the advent of electric typewriters, some typists were typing in excess of 80 WPM, but they could not easily make corrections. Correction methods now are designed into the typewriter. Some typewriters even have small screens, so you may edit your work after keypunching and before printing begins.

With word processing, not only are a few lines of text edited on the screen, but whole pages are edited at one time. Sentences or whole paragraphs can be moved, or "cut and pasted," and large amounts of work may be compactly stored on disk for

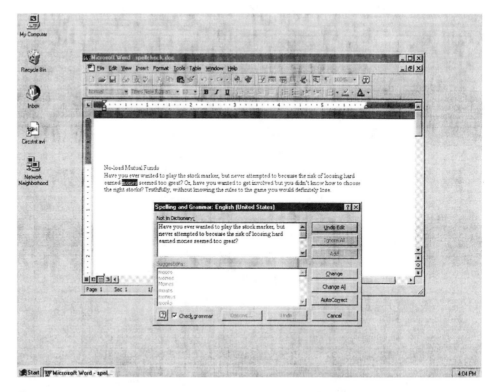

Figure 8.19   A screen printout of a spell-checker at work. The word *mones* is highlighted and possible replacements are offered in the suggestions window below.

future editing and/or integration into new documentation. Word-processing text may also be sent to other computers via modem linked with a telephone system.

The latest word-processing packages contain **spelling checker** features (Figure 8.19), dictionaries, and a thesaurus. One reason for many to learn word processing is the "spell-check" feature. There is no excuse for poor spelling when a spell-check program can highlight all the misspelled words in a document in seconds. A built-in thesaurus yields alternate word choices for words you select. More advantages can be listed, but you must try one or two different types of word-processing software before you can appreciate any advantages.

EXAMPLE
8.6

You have just completed a report on the stock market for a communications class. Proceed to spell-check the document and allow the word-processing software to make the corrections.

Solution   The first portion of the composed text, with the word *mones* highlighted, is at the top of the screen (see Figure 8.19). Below this screen is the spell-check window opened to display suggested replacements for *mones* (moneys or monies are correct spellings found in the word processor's dictionary).

If the misspelled word has only one similar word in the dictionary, the spelling checker could give you only one—the correct—selection. Alternatively, the word may be so badly misspelled the software would offer no alternative. However, the word would still be indicated as misspelled.

Did you notice that the word *loosing* was not highlighted? Even though it is used wrong, it *is* a word found in the word processor's dictionary. Misused words are not edited by the checker.

You can add to the word processor's dictionary. User-created dictionaries are known as **lexicons** or **alternate dictionaries.** Engineering lexicons are available in commercial software and contain technical words not found in standard dictionaries (e.g., the word *four-way* [valve], as used in hydraulics). Technical software is also available to create equations using the Greek alphabet and other symbols used in science and technology. The equations can then be blended neatly with standard text.

Do not become a victim of learning one word-processing package without trying out new ones as they become available. More powerful and more specialized word-processing software packages are introduced each year. Refusing to switch, once having learned one word-processing program, will cause you to become outdated and less productive.

After you have written a document you will wish to save it. To do so, you must create a file. The file must have a name—FILENAME—of up to eight characters (more than eight characters can be used with Windows 95); it is convenient to assign an extension also. Extensions are separated from the file name by a period. **Extensions** are up to three characters long and are optional, but help to identify whether the file is a basic program—.BAS; a program stored on disk in a ready-to-use form—.COM or .EXE; a spreadsheet calculation file—.CAL; or a word-processing document—.DOC. An example of a word-processing file name with extension is

FILENAME.DOC

It must be emphasized how important it is to begin to build your competency with word-processing applications as soon as possible. The first time you discover mistakes in a report that you have saved on disk, and find how easy it is to correct the mistakes and reprint the report, will make you a "true believer" in this most useful type of application software.

Electronic Spreadsheets

Spreadsheet application programs produce a display of rows and columns that you can insert data into or write equations to. Imagine the accountant's ledger. Filling the ledger out with pencil takes some time, and the result isn't easily read by others. As new information comes in, the ledger or spreadsheet must be modified. More importantly, the calculations that yield final information—such as the "bottom line"—must be computed over and over again.

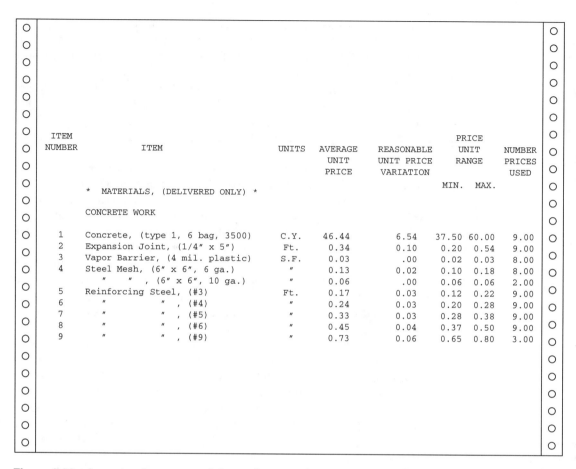

ITEM NUMBER	ITEM	UNITS	AVERAGE UNIT PRICE	REASONABLE UNIT PRICE VARIATION	PRICE UNIT RANGE		NUMBER PRICES USED
					MIN.	MAX.	
	* MATERIALS, (DELIVERED ONLY) *						
	CONCRETE WORK						
1	Concrete, (type 1, 6 bag, 3500)	C.Y.	46.44	6.54	37.50	60.00	9.00
2	Expansion Joint, (1/4" x 5")	Ft.	0.34	0.10	0.20	0.54	9.00
3	Vapor Barrier, (4 mil. plastic)	S.F.	0.03	.00	0.02	0.03	8.00
4	Steel Mesh, (6" x 6", 6 ga.)	"	0.13	0.02	0.10	0.18	8.00
	" " , (6" x 6", 10 ga.)	"	0.06	.00	0.06	0.06	2.00
5	Reinforcing Steel, (#3)	Ft.	0.17	0.03	0.12	0.22	9.00
6	" " , (#4)	"	0.24	0.03	0.20	0.28	9.00
7	" " , (#5)	"	0.33	0.03	0.28	0.38	9.00
8	" " , (#6)	"	0.45	0.04	0.37	0.50	9.00
9	" " , (#9)	"	0.73	0.06	0.65	0.80	3.00

Figure 8.20 A construction company's "cost of concrete" survey of nine surrounding counties resulted in the spreadsheet calculations of average price, reasonable variation (using the standard-deviation calculation), and the minimum and maximum expected costs.

As an example, the screen in Figure 8.20 demonstrates a good use for a spreadsheet at a local construction company. The spreadsheet has organized the results of a survey of construction material costs. Not shown are the figures on "concrete costs" collected from the surrounding nine counties. These figures were input to the computer in rows on the spreadsheet immediately to the left of each cost item.

Reviewing the column headers we see item number, item, units used, average unit price, reasonable unit price variation, the price range calculated from the variation column, and the number of counties surveyed (in the number prices used column).

We will examine item number 1 (concrete) to further clarify the spreadsheet results. The cement used to make the concrete is Type 1; this specifies some properties expected, such as 5 percent air entrainment and 2500 PSI compressive strength seven days after pour. It takes six bags to make one cubic yard (see units column [C.Y.]), and the final compressive strength will be 3500 PSI.

After the survey of suppliers has been completed and the data entered, the average unit price (for six bags) is calculated at $46.44. The spreadsheet software also contains standard-deviation (one measure of variability) equations and calculates the variance of the nine-county survey. The result is the "reasonable unit price variation." Specified multiples of the variation are used to gain the maximum and minimum costs for the concrete needed for a construction job in units of cubic yards.

The real value of the spreadsheet will be seen when the next survey is conducted. After the collected information is entered into the computer, the new minimum and maximum costs will be automatically calculated.

Specialized spreadsheet programs available to technicians automatically calculate such varied parameters as cylinder size, project cash flow, and spring design. These specialized spreadsheets offer predesigned templates. The *template* contains headers, equations, and all other information not changed on the spreadsheet as data are entered.

Perhaps the greatest power of the spreadsheet is to play "what if?". Data can be changed to see what effect is created on the bottom line. This is of great help to entrepreneurs planning future strategies.

Data Base-Management Systems

A data base is information collected and organized in a certain way. Data base-management systems allow you to enter, sort, update, find, and print reports of data quickly and easily.

A college's schedule of classes for a certain quarter is a data base you are familiar with. Each line of the schedule of classes is known as a record. Each of the elements within the record (i.e., course number, classroom number, meeting time, etc.) is known as a field. Fields, in other words, make up a record. The schedule of classes is indexed by course number. The biggest advantage of the data base-management system is its ability to index on certain fields. For instance, you could check for course conflicts in the engineering technology course sequence by selecting courses only in the engineering technologies and indexing to meeting time.

In industry, large data bases are maintained on manufacturing processes. For instance, in a metal-casting operation it is important to know such things as pour temperature, carbon content, and length of time to pour. Today's powerful PCs can be connected to accept data from manufacturing processes, and the data base can organize millions of data points in ways that can be easily interpreted by the technologist. The ability to produce charts and graphs is an integral part of most integrated data-base packages.

Simulation Software

Application software that can simulate real-world objects is known as **simulation software.** Examples of applications range from computer models of weather patterns, to stress analysis on machine tool parts, and to the flight characteristics of an

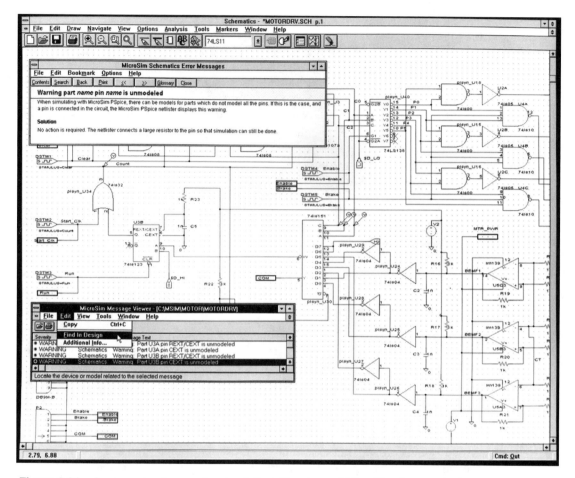

Figure 8.21   An example of computer simulation. (Courtesy of MicroSim Corporation)

airplane. Most simulation software requires complex mathematical calculations, which, in turn, requires supercomputers.

One popular simulator that is used in many schools and colleges is PSpice®. **Spice** is an acronym that stands for *Simulation Program with Integrated Circuit Emphasis.* It was developed to simulate electrical (integrated) circuits before they were built. PSpice® is a commercial version of the Spice family of software that works on a PC. Figure 8.21 is a computer screen from PSpice®. After the circuit is constructed the simulation software can troubleshoot the design and show output signals for selected input signals.

Before proceeding to the next section, *you should try to operate at least one of the four types of application programs described above.* Technicians and technologists often need to develop and modify programs and even write their own programs (Chapter 9). But, most of your career will be spent using preprogrammed applications like those described in this section.

8.4 Internet Guide

The **Internet** (or the Net) is an international computer network. A **network** is formed whenever two or more computers are connected so they may communicate (see Figure 9.9). A **Local Area Network** (LAN) is a network at one location such as a manufacturing facility or your college. Each computer connection in a LAN is called a **node.** Networks that cover wide geographic areas are known as **Wide Area Networks** (WAN). WANs connect entire states or countries. The Internet connects the established WANs around the world. The Internet began its existence in the U.S. (in 1969) as a method of maintaining digital (computer) communication in case of nuclear attack. Today, it is used by anyone wishing to communicate with the world.

The Net will definitely be a part of your career in technology. In the past, technologists had to rely on the nearest technical library to find answers to the questions that continually arise. Information was seriously limited and *reinventing the wheel—* or laboriously developing a product, part, or process that had already been done by someone else—was a common practice. With the Net you will be able to keep up without even leaving your office. Information that was costly in time and energy has suddenly become a bargain.

In addition, *the Net has created thousands of technical jobs.* LANs must be developed at local sites. **Servers,** high-speed computers sharing programs and data files on networks, need to be serviced. A multitude of other equipment making up the Internet will need servicing as well. The Net is growing at a rapid rate, leading to new equipment installations, software development and installation, and technical support to users.

Your PC should have at least an Intel-based Pentium or Macintosh 68040 microprocessor to use the Net efficiently. To access the Net you need a modem and an Internet access provider. The term **modem** stands for modulator/demodulator. Modulation converts the computer's digital pulses into audio frequencies and demodulation converts them back into pulses for the computer. At this writing the fastest modems run at 56,600 bits per second (56.6 kbps). It usually pays to purchase the fastest model available. Eight megabytes of memory (RAM) is the minimum needed for efficiently using the Net, with up to 32 megabytes recommended. The most popular online services such as America Online (AOL) offer access, or you may choose to go with an independent or local Internet access provider. *It is important to make sure your provider offers a local phone line and technical support.* (Net-based technology is moving very fast and by the time you read this section of the text we may all be online with WebTv. In 1995 Sony introduced its WebTV Internet terminal allowing you to surf the Net through your home TV.)

In place with every Internet starter package is the **Transmission Control Protocol/Internet Protocol** (TCP/IP) software that provides the language all computers on the Net use. The TCP protocol controls the transfer of data and the IP protocol provides routing instructions. To use the Net your IP address and the address of the provider's server (the shared computer in the network) must be known. A typical Internet address is

NAME@COMPANY.COM.

One of the most used benefits of being online is e-mail. There is no cost beyond the subscriber fee paid each month. **E-mail,** or electronic mail service, allows you to instantaneously reach anyone in the world connected to the Net. E-mail addresses are easily kept in files, so sending a business message takes no more effort than typing the message and downloading the particular address from your personal-address directory. You can also attach word processing, sound, or graphics to an e-mail. Compared to land-based mail systems (snail mail), there is just no comparison in time and cost savings.

To download information, you will need a **File Transfer Protocol** (FTP) program. Many times FTP will be part of other Net software (see browsers, below), but separate programs will often be more efficient. Many of the tools you need for the Internet may be downloaded from the Net itself. File transfers often take a long time. For instance, a modem running at 14.4-kbps takes about one second to download each kilobyte of information. If a transfer is interrupted, all of the information will be lost and the whole process must be started again. Make sure that other phone extensions in your home are not to be used during downloads and hope that there are no power interruptions.

To get around on the Net you will need some form of tool that can locate the information you need. **Gopher** is an early menu-based system for finding information by subject or by geographical region. The term itself came from the words "go for." There are more than 1300 gopher sites around the world, most of them associated with universities. One popular site used to find out more about gophers is at gopher://gopher.tc.umn.edu:70/. The site is maintained by the University of Minnesota. Whenever you use Gopher you are using a service called Telnet. **Telnet** connects you to the host computer you are accessing data from. For example, to access the New Jersey Institute of Technology, you key in: telnet www.njit.edu. There are other tools used to allow the Internet to work. It is really not necessary to list all of them since most are incorporated in the provider's software or the latest browsers (see below).

The heart of the Internet and the most user-friendly is the **World Wide Web (WWW).** The language of the Web is hypertext markup language, or HTML code. **HTML** code is used to make up the graphics and text of a web page. Each Web page is hyper-linked, or connected, as in a *web.* A **hyperlink** is a part of a web page that may be selected (usually by clicking on it with a mouse) that links it automatically to other information. The new information may be at the same location or thousands of miles away!

The latest generation of graphical-user-interface software used to get around the Web is known as a **browser.** Browsers incorporate all of the tools mentioned previously, such as FTP and Telnet. As Gopher is to the Net, browsers are to the Web. You can access Gopher, as well as other Internet addresses such as FTP and Telnet, from your Web browser.

The two most popular browsers are Netscape Navigator (Figure 8.22) and Microsoft Internet Explorer. This type of software is where you start when you want to access the Web. Referring to Figure 8.22, you will find a dialog box used to enter the Web address you wish to access. A web address is known as a **Uniform Resource Locator (URL).**

When on the Net you need other software to help you find what is needed. The software used in your search is referred to as a **search engine.** To search for information in the Gopher system, use Archie. Archie software has indexed all of the public domain files on the Net. Archie searches the files for you and lists locations and file names, but it will not retrieve the file. To retrieve it, use ftp or a browser.

Figure 8.22 A popular web browser.

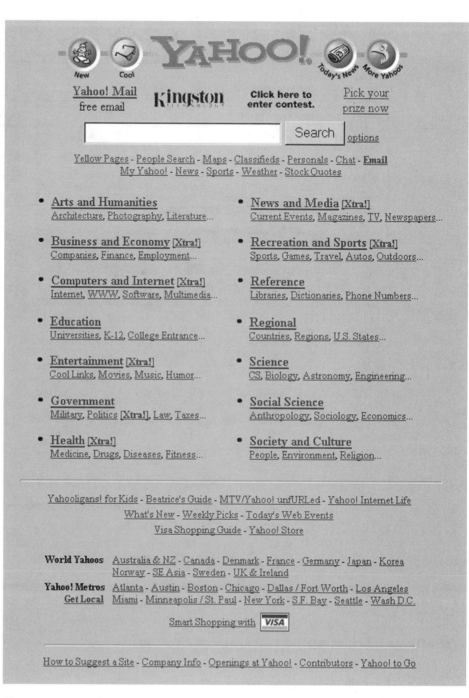

Figure 8.23 A popular search engine.

Search engines for the Web come in two basic forms: people-constructed indexes of Web sites like Yahoo (Figure 8.23), and robotically constructed information contained in Web sites such as Altavista. People-constructed indexes are often referred to as directory sites. A **directory site** lists broad topic areas and offers a good place to begin a search for information on a particular topic. A true search engine sends out "robots," or automated Web browsers, that surf the text on the web continuously to find any reference to a given search topic. The advantage to using a true search engine rather than a directory is that the depth of the search is greatly enhanced. The disadvantage is that a search may retrieve thousands of "hits," many more than you would have time to read. To use any search engine you should always check out the Frequently Asked Questions (FAQs) at the site. Reading the FAQs can save you hours of frustration by letting you know how the search strings should be constructed.

The Net is continuously evolving. The software needed—not listed here— ranges from those programs that stuff files to those that protect your computer from viruses. Look at it as you would any other technology: a continuous, neverending learning experience.

8.5 Computer-Aided Design (CAD)

Just as word processing eliminated the problem of correcting mistakes for the report writer, computer-aided design (CAD) software eliminates the problem of correcting mistakes on the drawing board. It is much easier to change data on a disk than to erase and recopy on a hard-copy drawing.

The drawing information, edited on a CAD system, may be printed any time after a change is made. Keeping documentation on disk is far superior to keeping it on hard copy, which must be stored in expensive flat files. Also, as with other software packages, the information on computer can be electronically relayed to other departments—even sent to other plants.

CAD software does more than give the operator the ability to construct technical drawings on a computer. CAD systems also dimension parts, keep track of material needs to construct the part, and even check for interference between parts on working drawings.

The PC has only recently become powerful and fast enough to handle CAD applications. Prior to the early 1980s, only mainframes or minicomputers could handle the relatively large amount of processing required by CAD-application software. Until microcomputers could operate the software, smaller companies could not afford to implement CAD. Today, system costs are dramatically lower, and productivity increases of 5:1 are common. CAD systems will, in the near future, be used in virtually all manufacturing operations. (Productivity may decrease temporarily as operators are trained on new systems. It is naive to believe a 5:1 productivity gain can be accomplished immediately after purchasing a system.)

With all of these advantages, you are probably ready to throw away your drafting table. Don't! Almost all drafting instructors feel you must begin with "board work." You need to possess a thorough understanding of the language of drafting before you

Figure 8.24 Most technical colleges offer required or elective courses in CAD, but only after the student has completed a traditional "board" course. (Courtesy Central Ohio Technical College)

can become efficient on a CAD system. Technical colleges universally require drafting in the first quarter of an engineering technology curriculum. Although some CAD may be integrated in the first course in drafting, your primary CAD course should follow only after you are competent with traditional drafting tools (Figure 8.24).

Minimum system requirements for most CAD applications include

1. at least 32 MB (megabytes) of RAM,
2. provision for connecting (interfacing) additional input devices (e.g., digitizers, light pens, digitizer tablets), and
3. connectivity for additional output devices (e.g., plotters and higher resolution monitors).

See Figure 8.25.

Many CAD vendors offer **turnkey systems.** Turnkey systems come with all of the compatible peripherals and supporting software you need to perform your application. Much time is saved and fewer disappointments encountered during startup when a turnkey system is purchased. The cost is usually not much more than when individual components are purchased.

Figure 8.26 is a typical CAD workstation. The student is using two monitors, one for the drawing and one for keeping track of input commands to the system and the system's responses (text/data monitor). The two inputs she is using consist of a keyboard and a digitizer (with tablet).

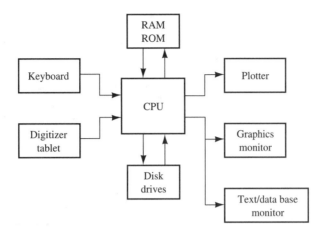

Figure 8.25   A block diagram of a turnkey CAD system.

Figure 8.26   How many input and output devices can be seen in this photograph?
(Courtesy Central Ohio Technical College)

Determining what peripherals are necessary can be simplified by determining specifications required. For instance, in purchasing a monitor the number, or density, of **pixels** is a good guideline (Figure 8.27). The number of pixels on a CAD monitor should be equal to or greater than 1024 × 768. High-resolution monitors used in industry have

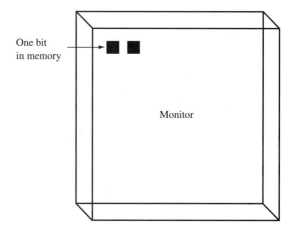

One bit
in memory

Monitor

Figure 8.27 Each dot on the computer monitor is controlled by at least one "bit" in memory. Each dot is referred to as a pil (black and white only) or a pixel (color).

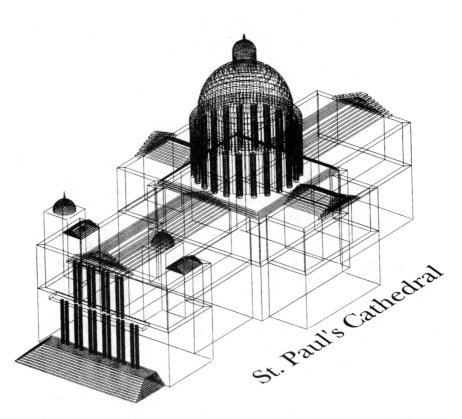

St. Paul's Cathedral

Figure 8.28 The output of a large pen plotter allows the technician great detail. (Courtesy Autodesk, Inc.)

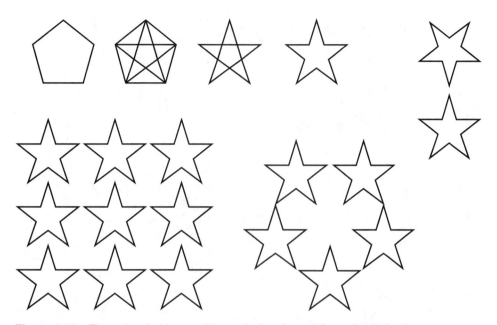

Figure 8.29 The only primitive used in producing the star formation is the line.

a screen pixel count of up to 1200 × 1200. Higher pixel density makes smoother curves, improving overall picture quality.

The **plotter** represents another important output device. The plotter makes hard copy of the product represented by digital information in the computer. (A screen or monitor drawing is sometimes referred to as *soft copy.*) Most CAD systems use a pen plotter (Figure 8.28). As the pen moves along one axis (i.e., *X*-axis), the paper is moved back and forth along the other (or *Y*-axis). Industrial systems usually invest in a multipen device, capable of producing multicolored drawings. A wide variety of plotters is available. For instance, the 3-D plots, in Figure 8.30 were done on an ink jet plotter.

Input devices, or **digitizers,** are transducers or encoders. That is, they take a key press at a certain location (as on a digitizer tablet) and convert the mechanical information to electrical information. The *X-Y* coordinates of the digitizer are converted to digital information signals. Light pens, trackballs, joysticks, and the "mouse" (which does not require a digitizer tablet) represent other input transducers. To evaluate input or output devices you must try the different types. Your two-year college is a great place to try out the many types of peripherals available.

Familiar now with CAD equipment, we will examine how the CAD system is used. The basic drawing shapes the computer automatically processes are known as **primitives.** Lines, circles, arcs, ellipses, curves, points, and text are primitives found in most CAD software.

One primitive—along with some commands like SNAP, ERASE, TRIM, and MIRROR—is used to produce the stars in Figure 8.29. Only the line primitive is used.

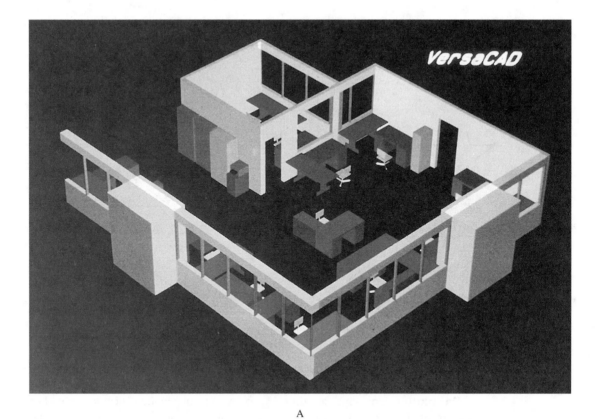

A

B

Figure 8.30 Architects may use 3-D visualizations in presentations to clients. (Courtesy VersaCAD [A] and Autodesk, Inc. [B])

View clockwise from the top left. First, the pentagon is produced by using a special command called POLYGON. (Some commands elicit queries from the software such as how long, how many sides, etc.) Using the line primitive and the SNAP command, the first star is then easily placed between the points of the pentagon. After the pentagon lines are removed by using the ERASE command, the star is completed by removing its inside lines. To remove inside lines the command is TRIM. Next, the MIRROR command is used to produce the two stars (note: they are mirror images). Finally the ARRAY command is used to come up with the final two patterns—use the "polar" option for the circular arrangement and the "rectangle" option for the rectangular array.

One other command might be mentioned before leaving this brief description of operating a CAD system; the HELP command is probably the most used command when using new software!

Creativity is the name of the game when operating CAD software. It is so easy to change drawings and view alternatives that "good" CAD operators can pursue more design possibilities while modifying the drawing on the screen and before printing the final drawing.

Before the development of the PC, most CAD applications were in the electronic design and drafting area. The circuit conventions and associated schematic symbols of electricity/electronics lend themselves to the CAD environment, and great productivity resulted from the marriage. High productivity makes the increased cost of a "mini" practical. Now, with microCAD and decreasing prices for hardware, the trend is to more applications in other engineering areas.

Engineering application areas with examples of how CAD is used in the representative areas are

- chemical engineering—batch process control;
- architectural engineering—design of commercial and private buildings and structural engineering (Figure 8.30);
- civil and construction engineering—plumbing, piping, and heating, ventilating, and air conditioning (HVAC);
- electrical and electronic engineering—circuit design and drafting, printed circuit board layout, and integrated circuit (IC) or chip design;
- mechanical engineering—product design, tool design, and drafting (Figure 8.31); and
- industrial engineering—plant layout and project management.

Computer-aided design and drafting (sometimes written as CADD) applications are more difficult to learn than the application programs we used in Section 8.3. Sometimes whole textbooks are devoted to one particular type of software product. You will want to schedule at least one course in CADD during your college education. Technologists must be aware of the value of this powerful application program.

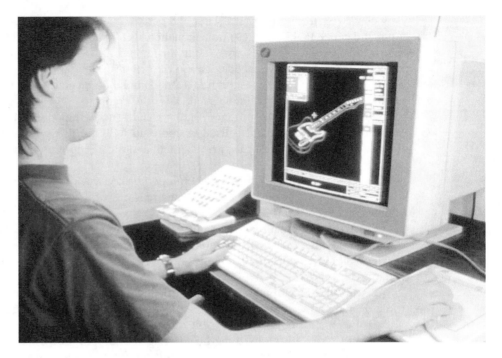

Figure 8.31 A mechanical engineering technologist designs a guitar. (Courtesy International Business Machines Corporation. Unauthorized use not permitted.)

8.6 Purchasing Your PC

As a new engineering technology student, you may wish to delay purchasing a computer, at least until after your first year. Most colleges have a variety of powerful computer systems. These computers are usually available in open laboratories on a day and evening basis. However, if you are considering the purchase of a PC, the following guidelines may be helpful.

Purchasing your PC can be challenging. First, there is the expense. The hardware for a minimal system includes the computer and monitor (usually sold separately). A printer may be necessary also, unless you depend on using the printer at your college. A computer with monitor could cost in excess of $1000. Software must also be purchased, but some software will come with the computer, and some technical software may be obtained from your college.

Your first decision should be what software you wish to use. The type of computer system you buy will depend on its intended use. For example, if you are enrolled in any of the technology fields that require drafting skills—civil, mechanical, or manufacturing—you will want to be able to design with CAD software. To run CAD software requires more memory, at least 32 MB of RAM and a monitor with

Figure 8.32   The Power PC chip is installed on a motherboard. (Courtesy of International Business Machines Corporation. Unauthorized use not permitted.)

higher resolution (such as a Super VGA monitor—1024 × 768 pixels), than would be needed for word processing and data base tasks.

Listed below are recommendations for buying your first system designed to operate technical software:

1. Selecting a microprocessor, or CPU—If your system is to run the latest software, choose the Motorola-based PowerPC (Figure 8.32) or the Intel-based Pentium running at 166 to 200 MHz. (Speeds of up to 300 MHz are now available.) Less-expensive computers will have chips made by companies other than Intel. The PowerPC is a newcomer and does not yet have all the available applications software that the Intel does. But the PowerPC has been accepted by both Apple and IBM as well as other hardware producers, which will mean increased software availability in the future.

2. Determining memory size—Be generous when ordering memory or RAM, especially for using the Web. Order at least 16 MB of RAM. To run CAD go up to 32 MB. Eight megabytes of RAM is the minimum required to run Windows 95 (see Integrated Operating Environments in Section 8.3). Also order a minimum external processor cache of 256 K, which is a standard feature of many systems.

3. Selecting the hard drive—Your new computer system should have at least a 1.6 GB (gigabyte) hard drive.

4. Deciding on a monitor—Most computers do not come with monitors. Select a 15 inch color monitor with a dot-pitch rating of .28 mm or less. The pitch value relates to the distance apart the characters may be on the screen. For CAD work you will need a resolution of at least 1024 × 768 pixels.

Figure 8.33 A wireless modem is attached to a laptop. (Courtesy IBM)

There are many add-ons you may consider, such as a fax-modem to communi-cate with other computers (Figure 8.33) and join the Internet (see Section 8.4 for more on modems and the Net), a CD-ROM drive (quad-speed or higher) to have large data bases available (e.g., the computer operating system, a U.S. telephone di-rectory, or a large encyclopedia), high-speed ports to speed up print jobs, and Energy Star compliance to lower electric bills and allow your computer to be on for late-night faxes.

If you are considering an add-on, you must be aware that there are a confus-ing number of bus and adapter technologies available today (also see Section 9.1, the data bus)—e.g., ISA and EISA buses and SCSI (pronounced "SKUH-zee") adapters. The bottom line is *do not buy a peripheral device or adapter card un-less it is compatible with your computer bus.* Your local computer dealer will know what peripherals you will need for your computer. A nice add-on for your new computer is a pair of USB ports. The ports will allow connectivity for peripherals designed in the future.

How about buying a used computer? There are many out there because many companies that use large numbers of computers must upgrade periodically. In addition, research has shown that most computers that have problems will have them early in their life, often within the first few months. This means that a used computer will rarely have problems later. If you are considering the purchase of a used computer, stay with the Intel-based 486 or above. Also before purchase, run the software you will wish to use, and make sure the RAM and the hard drive space is adequate. Make sure all of the necessary peripherals come with it and that there will be no hidden costs.

Glossary 8.2 contains computer terms used in Sections 8.2, 8.3, 8.4, and 8.5. Try to remember where and how they were used. Then, take time to reread sections with glossary items you are not sure of. Technologists should be familiar with these terms, and your instructor may hold you responsible for defining each of the following:

Glossary 8.2

ANALOG INPUT OR OUTPUT: A continuously changing, real-world signal (contrast earlier analog watches with today's digital watches)

BLOCK DIAGRAM: A representation of a system by blocking its individual parts—a great aid in troubleshooting

BYTE: A sequence of eight binary digits (bits); computer memory and disk storage space measured in kilobytes (KB) or megabytes (MB)

CAD(D): An acronym for computer-aided design; with the extra (D), CADD refers to computer-aided design and drafting

CHIP: An integrated circuit (IC); an extremely small electronic circuit with many elements and built on one "chip" of silicon

CIM: (Pronounced "SIM"); an acronym that stands for computer-integrated manufacturing

CISC: (Pronounced "SISC") Complex Instruction Set Computer; a microprocessor that runs large instruction sets at a high-level language

DATA BASE: A structure file of information, e.g., an address book or list of specifications for a certain product

DIGITIZER: A peripheral that converts analog to digital signals

DIGITIZER PAD: Registers the analog motion of a special stylus and converts the motion to digital information

DISK DRIVES: May be either hard or floppy (double density [DD] or high density [HD]); used to permanently store data in computer RAM

DOCUMENTATION: The supporting written material that comes with a computer system and peripherals (hardware), or with system and application software

EXTENSION: An optional part of a file name (e.g., FILENAME.DOC, in which DOC is the extension)

FIELD: The unit of information in a data base-management program; for an address record, a street address, city, or zip code

FILE: One collection of data existing in the form of a record to be stored on disk; files are named and accessed through disk directories

FLOW CHART: A graphical representation of the flow of information or electrical signals taking place in a system, in system software, or in both entities

FORMATTING: Preparing a new disk for use

IC: Initials for integrated circuit

IDE: Integrated Drive Electronics; a disk drive that contains its own controller and eliminates the need for an expansion slot

INTERFACE: The interconnection of two pieces of equipment so they will support each other

ISA: (Pronounced "ISUH") Industry Standard Architecture; the International Standard computer bus system used on many early PCs; EISA is a 32-bit extension of ISA

KEYBOARD: A common input device (encoder)

LEXICON: An application-specific (e.g., technical) dictionary added to the more general word-processing dictionary

LOOK-UP TABLE: A table in ROM accessible to the CPU so incoming information signals (inputs) may be compared with desirable parameters or unsafe parameters

MONITOR: A computer screen especially designed to handle the real-time aspects of, e.g., program timing—one form of output device

MS-DOS: The Microsoft disk operating system

NETWORKING: Connecting computer devices together so they can communicate with one another

PC-DOS: The IBM disk operating system

PCI: Peripheral Component Interconnect; called a mezzanine bus, and connects peripheral interconnect buses to the local (inside the computer bus)

PERIPHERAL: A computer's input or output device

PIXEL: A shortened term for *pic*ture *ele*ment

PIXEL DENSITY: The number of picture elements (pixels) per given area—higher pixel density means smoother curves, sharper corners, and overall better picture quality

PLOTTER: A device that produces hard copy of graphical data (the output of a CAD system)

PROGRAM: A sequence of instructions a computer follows to solve a problem. A complete program includes planning a solution, coding (specifying a set of computer instructions), and integrating the computer into the software (e.g., specifying printer formats).

PROMPT: An indicator on the computer display signaling that operator reaction is needed (e.g., A>)

RECORD: A collection of fields in a data base-management program

RISC: Reduced Instruction Set Computer (also see CISC); a microprocessor that runs smaller instruction sets at a lesser high-level language, making the chip a more efficient device than the CISC device

SCSI: (Pronounced "SKUH-zee"); Small Computer System Interface; an adapter or interface for connecting peripherals

Spell-check: Also known as spelling checker

Spreadsheet: An application program that manipulates values on a rectangular grid the user may input to

Template: A specially prepared spreadsheet with standard information or formulas that do not change

Transducer: A device that converts energy from one form to another

Turnkey system: A complete system of hardware, software, and (usually) service offered by one vendor

Word: An ordered set of characters handled as a unit by the processor

Word processor: An application software program that allows the entering, storing, retrieving, editing, and printing of text material

Write: To deliver data in RAM to a disk drive or other storage device

Problems

Section 8.1

1. List the three types of computers (see Table 8.1) and explore the features of each in one or two sentences.

2. Research the periodical indexes, especially the *Applied Science and Technology Index,* to find applications for each of the computers listed in Table 8.1.

3. Describe the differences between a minicomputer and a PC.

4. Define the computer by listing the three functions it performs. Draw a block diagram of a computer system containing the three functions.

Match each of the human characteristics in 5–10 with its related computer process.

	Human Characteristics	*Computer Process*
_____	5. hearing	a. memory
_____	6. recalling facts	b. processing
_____	7. moving an arm	c. input
_____	8. mentally adding two numbers	d. output
_____	9. speaking	
_____	10. seeing	

*11. Flowchart the registration process at your college. Compare your result with at least three other students' flowcharts. Even though they differ, explain how the differences might be analyzed to suggest improved registration procedures.

12. PCs are also known as microcomputers and possess a microprocessor. Are microprocessors only in PCs? Why or why not?

---

*Challenging problem.

Match the computer control functions in 13–17 with their logical input and output (two letters—one input and one output—match with each number).

	Control Function	Output	Input
_____	**13.** spark control (auto)	a. diaphragm control	A. radio transmitter
_____	**14.** camera	b. speaker	B. thermostat
_____	**15.** smoke alarm	c. gas valve	C. gas detector
_____	**16.** home heating system	d. timing circuit	D. photocell
_____	**17.** garage door opener	e. motor	E. manifold pressure

18. Describe the terms ROM and RAM. Explain how each contributes to the operation of a computer system.

19. Compare and contrast the terms *hardware* and *software.*

20. Accurately and succinctly define and use the terms in Glossary 8.1, in class and when examined by the instructor.

21. CPU stands for _____ _____ _____

Section 8.2

22. List the steps involved in the disk formatting procedure when using MS-DOS.

23. List the steps involved in the disk formatting procedure when using Windows.

24. Describe a GUI. Discuss in one paragraph the advantages of a GUI over using a disk operating system.

25. A popular dot matrix printer boasts a printing speed of 54 cps (characters per second). Assuming 60 characters per line and 25 lines per page, how long would the printer take to print a five-page document?

Section 8.3

26. Describe how application software differs from system software.

27. Research one type of application software and explain what it does in one paragraph.

28. Use one type of application software (e.g., complete one printed page). Copy the output to your personal disk. Submit the disk to your instructor for verification.

29. Describe all of the elements in row five for the spreadsheet in Figure 8.20.

30. Describe how an electronic spreadsheet works and give an example of its use.

31. Describe the functions or main features of a data base-management system. Give an example of its use.

32. Research one type of simulation software and explain what it does in one paragraph.

Section 8.4

33. In your own words, define a LAN and a WAN. Discuss the differences between the two.

34. How long will it take (in minutes) to download a one megabyte file, when using a 14.4-kbps modem?

35. Use a popular browser to find the currency exchange rate from the Japanese yen to the U.S. dollar. Submit the finding to your instructor.

36. Use a search engine to find employment information for your chosen technology. Submit the findings to your instructor.

Section 8.5

37. Differentiate between a low-resolution and high-resolution monitor in terms of number of pixels.

38. Obtain the brand name and cost for one type of CAD system at your college or at a local industry. Describe the system in terms of the requirements listed in Section 8.6.

39. What does the term *turnkey system* mean. Why would you purchase a turnkey system?

40. Accurately and succinctly define and use the terms in Glossary 8.2, in class and when examined by the instructor.

Section 8.6

41. Visit at least one local PC dealership. From its inventory, list the brand name and model number of a computer that will meet your needs as a technician. Be sure to use the specifications suggested in Section 8.6, Purchasing Your PC.

42. See Problem 41. After selecting your computer, determine what bus and adapter (interface) technology it uses and what peripherals you will buy. Obtain price data on the entire system including peripherals.

43. While at the PC dealership (Problem 41) discuss different computer operating systems (e.g., CISC vs. RISC systems).

Selected Readings

BYTE. Peterboro, N.H.: McGraw-Hill Inc. Published monthly.

Freedman, Alan. *The Computer Glossary*, 6th edition. New York: AMACOM, a division of the American Management Association, 1995.

Internet World. Westport, Conn.: Mecklermedia Corp. Published monthly.

Kennedy, Angus J. *Rough Guide to the Internet*, Version 2. London: Penguin Group, 1996.

MacUser. Belmont, Calif.: Ziff-Davis Publishing Company. Published monthly.

PC Magazine. New York: Ziff-Davis Publishing Company. Published biweekly.

User's manuals for application software products you use with this chapter.

9

Programming and Industrial Automation

This chapter will focus on the use of computers in industry. The need for technicians who can use computers in business and industry cannot be exaggerated. Millions of dollars are spent each year to train personnel in computer skills. Modern industries are not at all like those of even ten years ago. For example, one major automobile manufacturer describes a new slotting and shearing plant in this way: "Three-hundred-thousand tons of steel will be shaped each year into fenders by 38 employees. Nobody will touch the steel. All the people will do is handle the computers."

As this example makes clear, repetitive physical labor is being reduced. The new worker will deal with abstractions, often symbolically represented on a computer screen.

A sample of continuing education course topics, that might appeal to technologists working in today's industries, include

- programming in *Structured BASIC, C, PASCAL, ADA,* and other computer languages (see Appendix E),
- applying *p*rogrammable *l*ogic *c*ontrollers (PLCs) in industry,
- applying personal computers (PCs) in industry,
- using specific operating systems (e.g., UNIX, DOS, OS/2, Windows),
- designing and connecting *l*ocal *a*rea *n*etworks (LANs),
- installing MAP/TOP communication systems,
- maintaining "smart sensors,"
- troubleshooting specific computer systems,
- installing and maintaining *c*omputer *i*ntegrated *m*anufacturing (CIM) systems, and
- installing and maintaining advanced manufacturing systems, or *f*lexible *m*anufacturing *s*ystems (FMSs). (See Figure 9.1).

In this chapter we will first consider how to interact with the computer more directly by writing computer programs. Then, we will consider how the computer is used to control processes in industry.

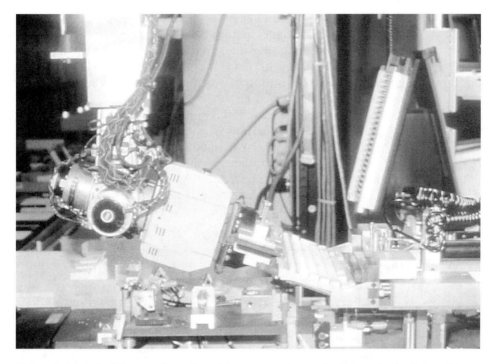

Figure 9.1 A flexible manufacturing system can make products to order for the customer. (Courtesy International Business Machines Corporation. Unauthorized use not permitted.)

9.1 Programming Languages

As a technician you may be required to write your own user-specific software to customize computer systems to perform specific tasks—e.g., control machines or processes and collect data. Like the application developer, you will need to work with a programming language.

Any computer language must be understood by both the computer and the programmer. **High-level** computer languages resemble the languages of humans and use familiar words and phrases. Each human language, regardless of its origin, has rules for spelling and grammar. The orderly arrangement of words and phrases is governed by the rules of **syntax.**

A computer also recognizes syntax, and much more rigorously than any human does. One of the messages the computer will give you when programming will be

Syntax error

which means the computer does not understand what you have typed in. The computer is indeed a "dumb machine." It understands only commands and instructions it has been programmed to accept.

Learning a computer language is like learning a foreign language, but surprisingly, you may find it easier to learn. The rigorous approach necessary with computer languages is refreshing after attempting to learn a foreign language with its many variations from one geographic area to another. However, true competency in any computer language takes place only when the programmer has the equipment and time to practice, practice, practice.

The only language the central processing unit (CPU) understands is *machine language,* a language consisting of only ones and zeros. The two symbols programmers call ones and zeros may be thought of as ON and OFF, YES and NO, VOLTAGE and NO VOLTAGE (the actual electrical signals within the computer circuitry), or any other pair of opposite conditions.

No matter how complex any decision is, it can be made by a series of YES or NO responses, as the following example demonstrates.

EXAMPLE 9.1

Out for a drive, you see a service station ahead on the left. Decide, by using YES or NO responses, whether or not to stop for gas.

Solution   You frame the following questions, then answer them with a simple YES or NO.

Question	YES or NO
Do I need gas?	YES
Is the brand name acceptable?	YES
Do I have enough money with me?	YES
Can I make a left turn here?	YES
Can I make it to the next station?	NO

Some of these questions did not have clear-cut answers but required decisions on your part. However, given these responses, you decide to stop for gas.

Of course the scenario in the example could be expanded to more complex questions, and decisions can still be made using only one of two responses. Our mind frames and decides questions quite fast, but the computer is faster still. Once a program is written to ask the right questions, the computer will react (think?) faster than the human. (Of course the computer does not think, but is programmed to respond in specific ways. Artificial intelligence programs designed to process information in much the same way we think are discussed further in Section 10.1.)

When a CPU accepts more than one YES or NO at a time, the number of possible combinations increases by a power to the base 2. In other words, the number of combinations equals 2^n, where

2 (the base)   = the number of states (0,1)

and n (the power) = the number of data lines processed at one time

Use Figure 9.2 to visualize the combination of states for a 1- to 4-bit CPU.

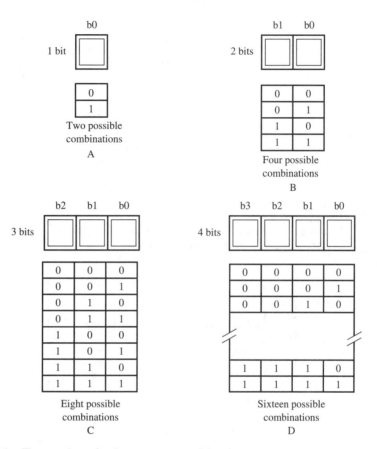

Figure 9.2 The number of unique states possible when a processor can process one, two, three, and four bits at one time.

EXAMPLE 9.2

A CPU is capable of processing three different data signals simultaneously. How many unique states are possible at one time?

Solution Using the equation, the number of unique states $= 2^n$, we find

$$n = 3 \text{ and } 2^3 = 8$$

Therefore, there are 8 unique states possible when three data lines are processed simultaneously (see Figure 9.2C).

One specification for a CPU is the number of bits it can handle at one time. In fact, CPUs are identified by this specification. When purchasing processors you will typically look for an 8-bit processor, a 16-bit processor, or perhaps a 32-bit processor. The higher the number of bits handled at one time, the faster and more powerful the processor.

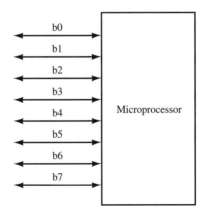

Figure 9.3 A block diagram of a microprocessor with an 8-bit data bus. The bus is bidirectional and accepts inputs or outputs.

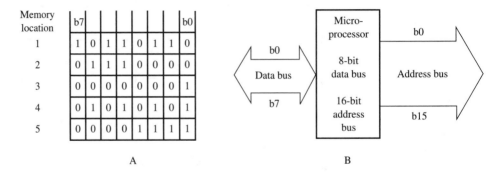

Figure 9.4 The address bus may transfer more bits than the CPU processes. (A) 8-bit data "words" stored in five consecutive memory locations. (B) A diagram of a microprocessor with an 8-bit data bus and a 16-bit address bus.

Buses, discussed in Section 8.5 and in Glossary 8.2, are also specified by bit count. A **bus** is simply a group of wires, optical fibers, or coaxial cable carrying data bits to and from components in the computer or to peripherals outside the computer. When you plug a board into one of the expansion slots in your personal computer, you're plugging into the bus. There are buses used to connect other buses. Eight-bit data buses carry 2^8 bits of information (or one *byte*) and are capable of conducting 256 unique pieces of information at one time (Figure 9.3).

Bus capacities also affect memory. If a microprocessor has an eight-bit data bus, then each memory location (address) must store at least eight bits of data (Figures 9.4A and 9.4B).

The most powerful PCs are 64-bit machines. But, regardless of the size, each of their CPUs understands only ones and zeros in the syntax (order) of that particular machine.

Programs etched into ROM are called **microprograms.** The built-in instruction set and the steps the instructions perform differ with each processor. Learning the instructions for any microprocessor would be a formidable task, especially if the technician or technologist tries to memorize the instructions as groups of ones and zeros.

For example, the instructions

01101000 = add two numbers

01100111 = subtract one number from another

represent two of many (e.g., up to 65 000) instructions that you would have to memorize.

Fortunately, there is a way to write programs without using all these ones and zeros. A high-level language may be used to communicate with the computer. **Assembly language** is one step up from machine language, using English words or abbreviations to construct commands. For example:

ADD—add two numbers together

STO—store the result in memory

CMP—compare the result with a code in a look-up table

BRE (BEQ)—branch to another instruction if they are equal

Assembly language words, or **mnemonics** (an alphabetical abbreviation used as a memory aid), will differ for each processor, but are, in all cases, easier to use than machine language. Assembly language mnemonics must be translated into machine language before the CPU can understand the instructions. Translation is accomplished by a language-translator program simply called an **assembler.** All high-level languages must be converted to the machine language the CPU understands (see BASIC assembler in Figure 9.5).

We learned that assembly language is one step higher than machine language, yet it is usually not considered a high-level language. In this chapter we will become familiar with one type of higher programming language—BASIC. The acronym BASIC comes from the words *b*eginners *a*ll-purpose *s*ymbolic *i*nstruction *c*ode. It is only one of many high-level languages used today (Appendix E). It happens to be one of the easiest to learn; hence, BASIC is the language most programmers learn first.

To convert the high-level language of BASIC to the low-level language of ones and zeros the computer understands, your computer must have a **BASIC interpreter** (or BASIC software). Fortunately, most—if not all—personal computers have a BASIC interpreter stored in memory (ROM). In Figure 9.5, the interpretation process flow is illustrated. **Note:** If a BASIC interpreter is not built into your PC, you will have to install or "LOAD" BASIC software from disk to RAM.

Your instructor will explain how to access the BASIC interpreter in your computer system. It usually requires a code such as **GWBASIC** or **BASICA.**

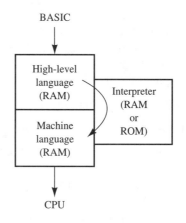

Figure 9.5 The interpreter converts a high-level language to the language of ones and zeros in the random access memory.

To confirm you have access to, or are in, BASIC try the command

CLS

After the return/enter key is pressed, and if you are in the BASIC mode, the computer will respond by clearing the screen and positioning the cursor (the blinking line where typing begins) at the top (under an OK) left-hand corner of the screen. (Function key commands could also appear across the bottom of the screen.) Read the following section while interacting with your computer system in the BASIC programming mode.

Finally, you must know how to exit the BASIC interpreter. For most operating systems the command is

SYSTEM

Using the PC as a Calculator

The **immediate mode** you enter when accessing the BASIC interpreter is sometimes called the **calculator mode.** You can begin your understanding of BASIC by using the computer as a calculator.

EXAMPLE 9.3   Use your version of BASIC to algebraically add the following sequence of numbers: $8 - 5 + 12 - 1 - 4$.

Solution   Begin all calculator calculations with the PRINT or ? command. Later, after writing some programs, you will see that without a command BASIC would read the first number (the 8 in this case) as a program line. We now type in

PRINT 8 − 5 + 12 − 1 − 4

and press the [RETURN/ENTER] key. The answer on the screen is

10

EXAMPLE 9.4

Using the calculator mode, calculate the following expressions: 12.8 × 36 and 12.8/36

Solution　We type in

? 12.8 * 36

(note the symbol used by BASIC for multiplication) and press [RETURN/ENTER], to see on the screen the answer

460.8

Next, we type

? 12.8/36 [RETURN/ENTER]

to see

.35555 __ __ __ __

Single-precision accuracy for many operating systems is seven or fewer digits—this can be modified by the programmer.

Just as the algebraic hierarchy was important to understand when using the calculator, it must be understood when using the computer. If you do not remember the order of operations, review Table 4.2, The Algebraic Hierarchy, in Chapter 4.

EXAMPLE 9.5

Calculate the following with paper and pencil and verify your answer with the computer's answer in the immediate mode.

3 × 4 + 4 × 5 + 5 × 6

Solution　The algebraic instructions are to first do the multiplications and then the additions. So, with paper and pencil you follow the steps

12 + 20 + 30 = 62

Using the computer, you type

PRINT 3 * 4 + 4 * 5 + 5 * 6 [RETURN/ENTER]

to see the answer:

62

Consider the following example:

$$\frac{5 + 4 - 5/6}{3 + 2}$$

With paper and pencil the numerator is first evaluated as

$$5 + 4 - 0.833 = 8.167$$

The sum in the denominator is then divided into 8.167, yielding the result

1.633

With the PC, as with the calculator, the problem must be structured with parentheses as

PRINT (5 + 4 − 5/6)/(3 + 2)

and yields the same answer as the paper-and-pencil solution.

Nested parentheses—remember those?—are handled by BASIC as explained in Chapter 4. BASIC will solve the parentheses from the innermost to the outermost.

EXAMPLE 9.6   Calculate the problem in Example 4.4, with the PC in the immediate mode.

Solution   The problem is

$$3[-2(4 + 7 - 3)] =$$

Just as we found with the calculator, BASIC uses only one form of parentheses and does not understand implied multiplication. You must type

? 3 * (−2 * (4 + 7 − 3)),

and the computer yields the correct answer

−48

With BASIC you will be introduced to a new power-of-ten format, but the use of such mathematical expressions remain the same as you learned in Chapter 4, Section 4.4.

The power-of-ten and scientific notation expression 9.84×10^7 is written in BASIC as 9.84E7; the small number 9.84×10^{-3} is written as 9.84E–3.

Note: BASIC does not use commas to represent large numbers. The number 4,300, for example, must be typed in as 4300.

EXAMPLE 9.7

Perform the following calculation:

$$(5.664 \times 10^5 \times 3.2 \times 10^{-13})/3.33 \times 10^{18} =$$

Solution   Press the following keys to enter the equation:

? (5.664E5 * 3.2E − 13)/3.33E18 [Return/Enter]

= 5.442883E−26, which is equivalent to 5.44×10^{-26}

BASIC handles a power in the following fashion:

2^8 is written as 2^8

and, as discussed in Chapter 4, the root is simply the reciprocal of the power, or

$$\sqrt[8]{2} = 2^{1/8}$$

Please refer to the example problems in Chapter 4 for further practice with powers of ten and powers and roots. Problems using the computer as a calculator are also offered at the end of this chapter.

The math operation symbols used in the calculator mode of BASIC are summarized below:

+   addition
−   subtraction
/   division
*   multiplication
^   power or (if reciprocal) root
       (SQR [SQR (4) = 2] is the square root function)
E   power of ten

BASIC offers other math functions, including trigonometric and logarithmic functions. The trigonometric functions as used in Chapter 6 are summarized as

SIN( )	sine
COS( )	cosine
TAN( )	tangent

The value (argument) the function operates on is written inside the parentheses.

The angles in BASIC are expressed in radian measure. Use $\pi/180$ to convert an angle in degrees to the same angle in radians (rad). To convert and find the SIN of a 45° angle, for example,

$$\frac{45 \text{ deg}}{1} \times \frac{\pi \text{ rad}}{180 \text{ deg}} = 0.785 \text{ rad}$$

and

$$\text{SIN}(45 * \pi/180) = 0.707$$

See Table 6.1.

9.2 Programming in BASIC

Technicians and technologists use application software most of the time, and they do not have to write programs. But one way to fully understand what causes the computer to do what it does—"to get into the computer's head"—is to write a program. Also, you can customize your computer's operating system by using BASIC programming.

You are well on the way to learning the BASIC language. By using the preceding section's examples for the calculator mode, you have already memorized the mathematics operation symbols. One of the reasons for your success is that the arithmetic symbols closely resemble symbols you were already familiar with. The same is true of the BASIC commands that follow. The developers of the BASIC computer language wanted programmers to easily understand the language. Hence, they used commands such as LET, PRINT, GOTO, and so on. See Appendix D for a list of beginning BASIC commands.

As an example, we will write a program to find the area and circumference for any size circle. See Chapter 6, Section 6.4, if you need to refer to the equations for the geometry of the circle.

You should enter this program line by line by following the example below. After entering the immediate or calculator mode (accessing the BASIC interpreter), enter a number that will become the first line number. Remember that while using the immediate mode you never began with a number, but used commands such as PRINT or ?. Had you written a number first, the computer would assume you are intending to write a program. This time, we intend to do just that.

Usually line numbers are used in steps of 100. That way, if you forget a line, it can be entered later in the right numerical sequence. In the following program, the letters PI indicate the value of the constant π (3.141 593). The program is written

```
100 CLS
200 INPUT "D =" ; D
300 LET PI = 3.141593
400 LET A = PI/4 * D^2
500 LET C = PI * D
600 PRINT "CIRCUMFERENCE ="; C
700 PRINT "AREA =" ; A
800 END
```

LET, at lines 300, 400, and 500, is always optional—try the program without the LET commands.

Before attempting to interpret each step, let's run the program first. To run the program, type the command RUN (there may be a special function key for this on your computer) and press [Return/Enter]. After successfully running the program a few times, and entering various diameters, you should be ready to examine the program steps, as follows.

Line numbers provide program flow from lowest to highest. If you wish to add a line, simply return to the program by typing LIST and type a number between the two line numbers where you wish to insert the line. The next time you LIST or RUN, your new line will be inserted in numerical order or executed in numerical sequence.

Program Line Analysis

Line 200	The INPUT statement stops the program and waits for you to enter something, a numerical variable in this case. The quotes are used to query the user of this new software. In this case, the diameter is what the program requires. If this were your own personal program, the query would not be needed, for you would know what is needed for the input.
Line 300	The first LET statement assigns the value of π (PI) to the variable.
Lines 400 and 500	The second and third LET statements set variables A and C equal to the equations used to calculate a circle's area and circumference. Two letters or a letter and a number may be used as variables in BASIC— e.g., A, AB, P4. All **variables** must begin with letters from the English alphabet. 4P, for example, is not a legal variable.
Lines 600 and 700	You recognize the PRINT statement used in the immediate mode. Again, the PC will type whatever you put in quotes and follow that statement with the value of the variable that follows the semicolon.

You may print out your program by entering the command LLIST. It is generally better not to use the PRINT SCREEN command to see your listed program.

If a program line is omitted or entered wrong you may not get anything on the screen when running the program. This is known as being "in a loop." You can exit a BASIC program at any time by pressing the two keys Control/Break (CTRL BREAK). See Table 8.2. *To exit the BASIC mode* you must also use a command statement. For many BASIC versions, the statement is SYSTEM.

The following example, with commentary, will continue to build your awareness of how the computer works with the data it is given. As always, it is best to use your computer to input and RUN each example.

A BASIC Program to Convert Vectors

In Chapter 6, Section 6.3, we converted vectors with magnitude and direction (polar form) to equivalent forms with X and Y components (rectangular form). We added the X and Y components of several vectors to solve for a resultant vector in the rectangular system (see Example 6.11). The resultant vector was converted to polar form, yielding its magnitude and direction (angle).

The following BASIC program will automatically perform conversions from polar to rectangular or from rectangular to polar. Write the program for your PC and use the 3-4-5 triangle (the magic triangle) and Example 6.11 to verify that the program works.

```
10 REM ' VECTOR CONVERSION - POLTOREC.BAS'
20 CLS : KEY OFF : PI = 3.141593
30 LOCATE 2,13
40 PRINT "THIS PROGRAM CONVERTS VECTORS FROM POLAR TO"
45 LOCATE 4,26
50 PRINT "RECTANGULAR OR FROM RECTANGULAR TO POLAR."
60 LOCATE 8,16
70 PRINT "ENTER '1' TO CONVERT FROM RECTANGULAR TO POLAR."
75 LOCATE 10,16
80 PRINT "ENTER '2' TO CONVERT FROM POLAR TO RECTANGULAR."
85 LOCATE 14,40
90 INPUT B
100 IF B=<0 OR B>2 THEN BEEP: GOTO 20
110 ON B GOSUB 300, 200
120 INPUT "DO ANOTHER? (Y OR N)" ; A$
130 IF A$ = "Y" OR A$ = "y" THEN 20
140 END
200 CLS : INPUT "VECTOR MAGNITUDE, V =" ; V
210 INPUT "VECTOR DIRECTION, ANGLE A =" ; A
220 X = V * COS(A * PI/180)
230 Y = V * SIN(A * PI/180)
240 PRINT "X =" ; X
250 PRINT "Y =" ; Y
260 PRINT: PRINT
270 RETURN
300 CLS : INPUT "X VALUE IS" ; W
310 INPUT "Y VALUE IS" ; Z
320 V = SQR (W^2 + Z^2)
330 PRINT "V =" ; V
340 PRINT "ANGLE A IS" ; ATN (Z/W) * (180/PI)
345 PRINT: PRINT
350 RETURN
```

After typing the program in, test it by using the 3-4-5 triangle in the beginning of Chapter 6. Type RUN and then select 1, choosing rectangular to polar conversion.

The program jumps to line 300 and asks for the *X* value. Use the smaller leg of the triangle as the *X* value—INPUT 3. When the program queries for the *Y* value, INPUT 4. The program responds by printing the resultant vector, the hypotenuse of the right triangle, as 5 (the result may require rounding). The program also shows angle *A*, the angle between the shorter leg (*X*) and the hypotenuse, to be approximately 53.1 degrees. Use your calculator to confirm the arc tangent (ATN) and verify the angle between the *X*-axis and the hypotenuse to be correct [$\tan^{-1}$ opp/adj or Y/X = angle *A* and calculator keypresses are 4 ÷ 3 = $\boxed{\tan^{-1}}$].

Note: Your calculator must be in the degree mode.

Next, the program returns to line 120 and queries "DO ANOTHER? (Y OR N)." The string (A$), also known as a **character string**, can contain any alphabetic or numeric character entered from the keyboard. In this case, A$ is defined as upper-case or lower-case y, and after typing y the program returns to line 20.

For further practice, select option 2 of your program to complete Table 6.2. The vectors are in polar form and the program will calculate the *X* and *Y* values of each vector. These values will correspond to the "movement east" (*X*) and "movement north" (*Y*) vectors previously solved with your calculator.

Program Line Analysis

10 REM	A remark is a note to the programmer; it is transparent to the program and appears only when the program is listed.
20 CLS : KEY OFF :	PI = 3.141593 The colon may be used to insert several operations on one line. The KEY OFF command turns the list of function key commands off at the bottom of your screen. Some operating systems may appear differently and not use this command. The two other commands on this line were explained previously.
30 LOCATE 2, 13	The LOCATE command is one of many screen **formatting commands** available to programmers. The LOCATE sets the next print statement to a row and column on the screen by LOCATE (row, column). There are 80 columns across the screen. The 2,13 in this case offsets the next printed line to row 2, column 13. A TAB command may be used with a PRINT command to offset a printed line by column only.
40 and 50	These two print statements inform the user of the software of what the program will do.

Program Line Analysis

70 and 80	These two print statements query the user of the program as to the input statement that follows on line 90.
100	This line protects you, the programmer, from an input other than what was asked for. The BEEP command is often used to warn a software user of an invalid input. Try using a value other than a 1 or a 2 to test this line of the program (B = <0 means B is equal to or less than zero).
110	The GOSUB command will skip to line 300 with a 1 and line 200 with a 2. The reverse order is keyed to the actual order of the conversion formats. If the 200 group of program lines were rectangular to polar conversion, the order on this line would be 200,300.
140	The command END does just that to the program when the user responds to line 130 with an N or n.
220	The argument in parentheses (A * PI/180) converts angle *A* in degrees to an angle in radian measure. BASIC uses SI units for angular measure (see Section 4.1, "Supplementary Units").
270 and 350	The RETURN statement works well with the GOSUB statement and always returns the program to the line immediately following the GOSUB line (line 110 in this case).

Certainly more comments could be made (see lines 45, 260, and 345 [45 is a line forgotten and added later; programmers use odd-numbered lines to indicate latest changes to a program]), but you can experiment with other lines by removing or modifying them. The flexibility of the computer allows you to experiment a great deal, and that is where the creative, problem-solving approach will be enhanced. Your instructor will recommend other programs to try and the Problems section will allow you still more practice.

BASIC was introduced in this chapter because it is one of the easiest high-level languages to learn. It is hoped this introduction helped you understand how a computer processes information. There are many other computer languages; Appendix E has a listing.

There are really two distinct categories of BASIC languages—structured and unstructured. In this chapter we have used **unstructured BASIC** to simply get a feel for a programming language. **Structured BASIC,** a newer form used in industry, sets off instructions into distinct blocks of operations rather than writing each line as it comes. Structured BASIC programmers often refer to unstructured BASIC as *spaghetti code.*

Also, it is important to realize that an interpreter is not the only way to convert a high-level language to machine language. You may also use a compiler. The compiler will run a program much faster than an interpreter.

The interpreter translates one line at a time. The **compiler** translates the whole program into machine language (ones and zeros) before running it. You will especially wish to compile your program if it contains graphic instructions; otherwise, you will wait all day for the interpreter to convert the program line-by-line and graph the programmed function.

QuickBasic Programming

The latest computers use QuickBASIC, or QBASIC. **QBASIC** is both a BASIC interpreter and compiler. Since the compiler translates a whole program at once, it runs programs much faster than the original GWBASIC. GWBASIC programs may be converted easily to QBASIC. QBASIC has pull-down menus and is mouse-supported, making your programming tasks much easier.

As an example of the use of QBASIC, use the GWBASIC program from the beginning of this section—the program to find the area and circumference of a circle. Use the following steps

1. Type QB (or QBasic) at the DOS prompt to initiate the programming environment.
2. Select **New Program** under the **File** pull-down menu.
3. Type in the program for finding the area and circumference of the circle. You do not need to type in the line numbers. Use the arrow, insert, and delete keys to correct any errors.
4. Select **Save** under the **File** menu. In the window that appears, name the program **CIRCLE.** The Basic extension **.Bas** will automatically be attached.
5. Use the **RUN** pull-down menu to execute the program. Put data in to test the finished program.
6. Press any key to return to the QBASIC main menu.
7. Compile your program by using the **RUN** menu and selecting **MAKE EXE FILE.** This makes an EXEcutable file that runs directly from the DOS prompt. Note! If you have not installed a QBasic compiler (QB4 or equivalent) you will not see the **MAKE EXE FILE** option under the **RUN** menu.
8. In the view window select YES to save your original program. In the following view window name the program CIRCLE.EXE. Choose either **MAKE EXE** (to try the program) or **MAKE EXE and Exit** (to run the program from DOS).
9. Run the compiled version of your program from DOS. You probably will not notice an increase of speed with the compiled version in this example, but longer programs with more calculations will run at least four times faster.

Welcome to the world of the programmer. Section 9.3 will explore how computers are used in industry.

9.3 Process Control

Computer-integrated manufacturing (CIM) represents the top of the factory automation pyramid (Figure 9.6). Before discussing CIM, however, we need to reflect on the systems leading up to the ultimate goal of full factory automation.

One of the most important industrial developments in the twentieth century was the idea of the control system having a feedback loop. The **feedback loop** returns process information (temperature, force, rotational speed) to the controller. You will eventually need to be familiar with a typical model of a control system, such as that shown in Figure 9.7A.

The basic control model involves a **controller,** (9.7B) an output to a prime mover (the electrical coils), a sensor, and the all-important feedback loop. The home heating system has an electrical controller that refers to a set point and switches larger currents to the heating coils when a lower than desired temperature is detected. The prime mover is the coils (and fins) that transform electrical energy to heat energy. The **sensor,** usually a bimetallic coil that expands and contracts with temperature change and opens and closes electrical contacts, senses temperature change and turns the controller on and off when the temperature of the set point is reached.

The sensor is important to technicians and technologists. Sensors used in industry are often **smart sensors.** These delicate and complex products have built-in

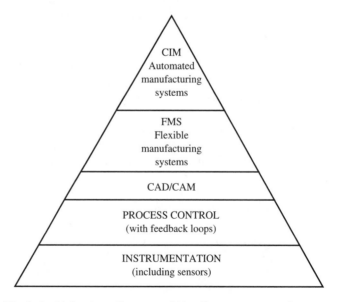

Figure 9.6 The industrial automation pyramid leading to computer-integrated manufacturing (CIM). The model begins with floor-level control and climaxes with total-plant control.

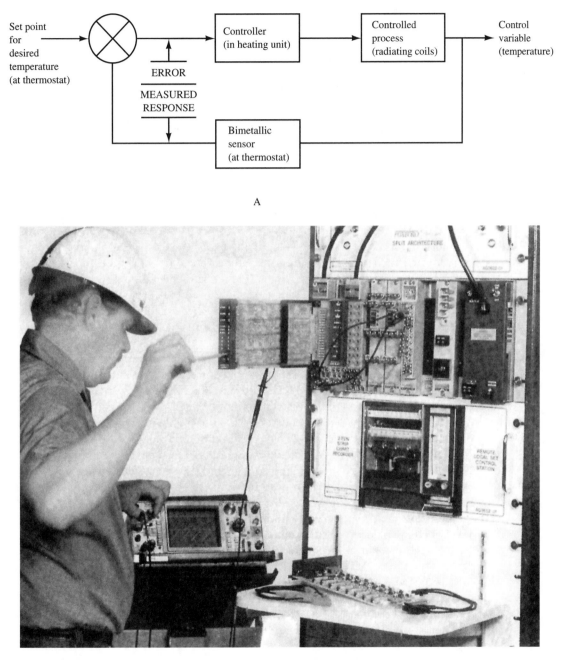

A

B

Figure 9.7 (A) The process control model for a home heating system has a feedback loop with a sensor. (B) Adjusting a process controller in industry. (Courtesy LAB-VOLT)

microcomputers along with microscopically small microsensors. Smart sensors exhibit the following capabilities:

- the ability to recognize errors in information and reject unwanted information and
- the ability to provide diagnostic data.

Smart sensors, located in dirty or electrically noisy environments, must be installed and maintained by knowledgeable maintenance personnel.

Numerical Control Leads to Computer-Integrated Manufacturing

Industrial computers were first used in the 1950s to control individual machine processes. The resulting system was called **numerical control,** or **NC.** NC is one of the most important industrial developments of our time. Before 1950, most products were formed and assembled using hard automation. **Hard automation** is built-in automation. In other words, the hardware, including the electronics, is built-in or hardwired. Hard automation cannot be changed easily and produces similar parts.

What if your company required parts of varying size, or only a few parts (a short run)? Until the advent of NC, changes and short runs resulted in parts that were prohibitively expensive.

You learned, in earlier chapters, how easily changes are made on a computer. With early NC, punched tapes were made to control machines such as lathes, milling machines, and drills. These tapes had to be modified as changes were made or as tapes wore out running through the machine controller. This was a minor inconvenience, however, and NC provided the short runs and the frequently changed individualized parts needed to respond to the diverse needs of the customer.

NC ushered in a new era, leading eventually to the completely automated factory. But the new technology was not greeted with enthusiasm by skilled machinists. These skilled workers worried about programmers, full of abstract knowledge, but possessing little or no practical knowledge. They feared computer programs would be written by these abstract programmers with machine steps such as "tapping a hole before it was drilled." Many machinists also thought they could not learn the new technology.

Today, machinists recognize the great value of the NC machine and the **Computer Numerical Control** (CNC) machine (with computer memory, requiring no punched tape—Figure 9.8). In fact, most skilled machinists have been introduced to the next step toward the automated factory, computer-*a*ided *d*esign and computer-*a*ided *m*anufacturing (CAD/CAM). You have already been introduced to CAD in Chapter 8. You know that design and drafting changes can be made easily, stored on disk, and later retrieved.

Computer-aided manufacturing (CAM) is simply a computer-controlled machine that communicates with a CAD system. When the final design's data from CAD are sent directly to a computer controller on the machine, the result is a computer-generated part from design to manufacturing, or **CAD/CAM.**

Figure 9.8 A technical student programming a CNC trainer. (Courtesy Central Ohio Technical College)

When computerized machine operations are connected, often to a main-frame or minicomputer, they are **networked** (Figure 9.9). In industry, the result is **computer-integrated manufacturing (CIM).** With CIM, communication in industry reaches an optimum level. Even while the designer is still designing the product, on the CAD system, the manufacturing engineers and technicians can be tooling up and verifying the manufacturability of the product. At the same time,

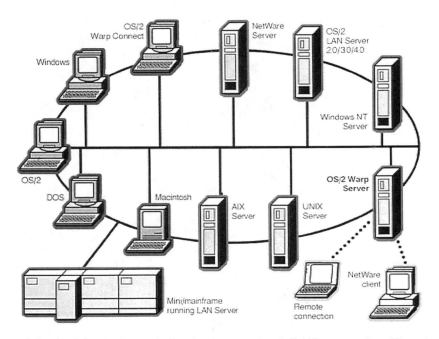

Figure 9.9 A minicomputer controls a local area network (LAN), connecting different operating systems to servers.

marketing personnel can be developing drawings of the product for promotional brochures.

A **flexible manufacturing system (FMS)** is a subset of CIM. An FMS is seen in sophisticated machine shops where short runs are common and part specifications are often changed. The system is composed of a group of machines in discrete work cells with automated material handling between machines. Activities are monitored and controlled by a multilayered computer system.

An example of an FMS system is a system used to repair jet aircraft engine turbine blades. The turbine blades are repaired regularly to maintain the efficiency and safety of the jet engine. Figure 9.10 shows a turbine blade and the turbine blade repair surfaces. The versatility of the FMS system allows each turbine blade to be treated individually. Prior to this system's development, each of the blades processed yearly was done by hand. The primary outcome of the implementation of the FMS is to increase productivity through

- improved metallurgical repair quality,
- improved dimensional repair quality,
- increased process yield by reducing scrap and rework,
- reduced consumable material consumption,
- increased production capacity, and
- decreased direct labor costs.

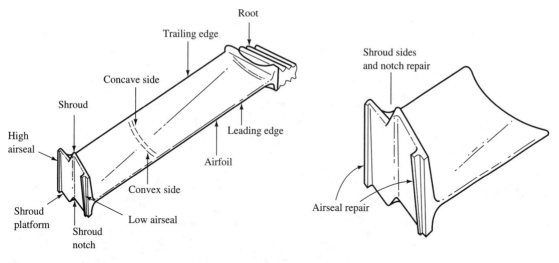

Figure 9.10 Turbine blade and repair surfaces.

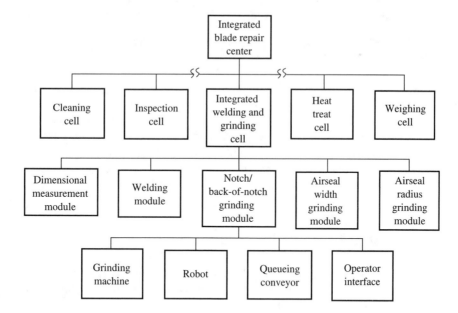

Figure 9.11 FMS distributed control system.

Activities within the FMS cell are coordinated and controlled by a distributed control system. A block diagram of a distributed control network supporting the FMS is diagrammed in Figure 9.11. Activities in the network include data processing, data management, and cell level process monitoring. Technicians are responsible for installing and maintaining the various controllers and communication networks within

Figure 9.12 This factory-automated system is designed to train technicians on automated systems. (Photo courtesy of Amatrol, Inc.)

the cell. Fiber-optic cables are used to connect equipment at the module level. Serial and parallel communications are used at the machine level.

Figure 9.12 shows a **factory automation system** used in many technical colleges to expose the student to hands-on applications. The system includes automatic storage/retrieval, laser barcode reader, and vision system.

Now that we have the big picture of industrial automation, we can appreciate two of the main components of the factory automation scheme—PLCs and PCs.

9.4 Programmable Logic Controllers, or PLCs

Working in concert with the powerful PC in industry is the programmable logic controller, or PLC. It is easier to appreciate the PLC by considering its electromechanical equivalent—the relay.

The **relay** consists of a coil and contacts that open circuits or complete them. When energized, the coil will electromagnetically open or close the contacts. If the relay points are pulled together, the relay is normally open (N.O.); if pulled apart, the contacts are normally closed (N.C.). It is easy to see the mechanical contacts must fail after 1000, 10 000, or 100 000 cycles.

The PLC replaces the older mechanical relays by logic gates in the computer. The PLCs are completely electronic—no moving parts. The result is a long list of major benefits, such as higher reliability (more cycles before failure), lower power

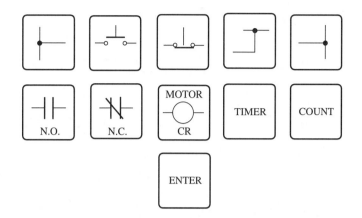

Figure 9.13 Keys found on a PLC's keyboard.

consumption, easier expandability, quicker modifications, reduced space considerations, and lower cost.

When technologists began replacing "hard automation" with rapidly improving computer systems, they had to develop a special logic that electrical maintenance technicians were familiar with. The older hardware components (relays) were connected in ladder diagram formats. Consequently, the logic for the PLC became **ladder logic,** making the transition for the technicians easier. Today's PLCs use many programming languages as well as the familiar ladder logic format.

Another advantage of the PLC is its design. It is specifically designed to be used in manufacturing. This means it is built to withstand harsh environments, is friendly to technical personnel, and has features that allow it to be easily connected to processes (highly efficient input/output [I/O] features).

Another advantage of the PLC is its **real-time control.** In other words, the PLC can be connected to collect data and control processes while the process is running. The PLC can usually respond faster (typically in milliseconds) to emergency situations such as emergency stops.

Figure 9.13 illustrates some of the keys on a programmable logic controller's keyboard. The keys depict the individual components used to construct a ladder diagram. Using keys such as these, the operator programs the PLC and creates a digital control that works the same as the circuitry shown in the ladder diagram in Figure 9.14.

EXAMPLE 9.8   Create a ladder diagram that will turn a motor on and off. Describe the diagram's initial conditions and the circuit operation from "start initiated" to "stop initiated."

Solution   See Figure 9.14. The **ladder diagram** is so named because it looks like a ladder when completed. Ladder diagrams have been used in industry for years to document the circuitry used to connect switches, relays, timers, motors, heaters, and many other components used in the manufacturing processes. The "rails" of the ladder represent

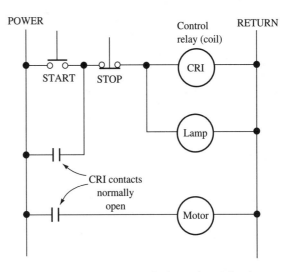

Figure 9.14 A ladder diagram depicting a typical start/stop circuit.

electrical power and electrical return; the "rungs" represent relay and component connections. The relays, of course, are replaced by logic instructions in the PLC.

A description of the circuit operation in Example 9.8 follows.

1. Initial conditions—All rungs are open-circuited, no current flows, and the motor in the last rung is off.
2. Press START—Pressing the start pushbutton activates the coil of CR1. The stop pushbutton is normally closed (N.C.) and the circuit is complete. Electrical current will energize the relay and the motor-run lamp connected in parallel.
3. Relay CR1 energized—When a relay is energized, the normally open contacts close and normally closed contacts open (no N.C. contacts are shown). One set of contacts for CR1 bypasses the start switch (connected in parallel). The start switch may be released and the circuit remains complete. CR1 is said to be "holding itself in."
4. THE MOTOR CIRCUIT—There is also a set of N.O. contacts in the bottom rung that close when CR1 is energized. Electrical current flows and energizes the motor.
5. Press STOP—The N.C. stop pushbutton opens the top rung, breaking the circuit, and de-energizing CR1. The CR1 relay contacts open, and the circuit is returned to its initial condition.

The PLC is relatively inexpensive and user friendly, easily interfaced with manufacturing machinery. It is a machine that many manufacturing technologists will come in contact with early in their careers.

9.5 Personal Computers (PCs)—User Friendly

Collecting data from manufacturing processes represents one of the earliest uses of the personal computer in industry. The PC can sample many characteristics—such as strength or temperature—every millisecond and store the data to be retrieved and evaluated later.

In statistical process-control operations, the PC is used to collect data (see Figure 9.15A), statistically chart the data (see "run chart"—Figure 9.15B), and let the operator know when the process is "in control" or "out of control." An "out-of-control" condition exists at point number 13 on the chart. The operator should stop the process or know what happened to the product during this period of production.

Other uses for the PC in industry include controlling inventory, designing (CAD), manufacturing (CAM), and controlling marketing and personnel data.

Today, the PC not only collects data, but also controls manufacturing processes. Easier to interface and offering higher bit counts and improved operating systems, the

A

Figure 9.15 (A) A portable PC, used as a data collector, can transfer its data to another, more powerful PC for data analysis. (Courtesy Central Ohio Technical College).

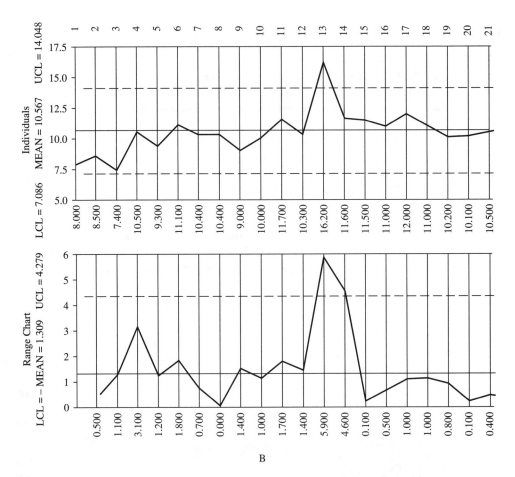

Figure 9.15 (B) The PC organizes the data, printing out "run charts" that tell machine operators whether their processes are producing parts in a stable or unstable fashion.

PC controls robotic systems (Figure 9.16) and many other manufacturing processes. The personal computer could be viewed as the user-friendly front-end to PLCs and distributive control systems in industry.

Many technical personnel feel that the PC will replace the dedicated (and less flexible) PLC. Other technologists argue that the PC's operating system has a difficult time responding as fast as the PLC. Current PC operating systems (e.g., DOS, UNIX, DOS/UNIX), without special design considerations, can do only one thing at a time.

Certain inputs—emergency stop pushbuttons, pump failure indicators, and safety alarms—must be answered very quickly. The real-time I/O in the programmable logic controller can respond faster in these situations. The PC may replace the PLC, but currently sales of programmable logic controllers are climbing and these more dedicated industrial machines have a firm foothold in distributed control systems.

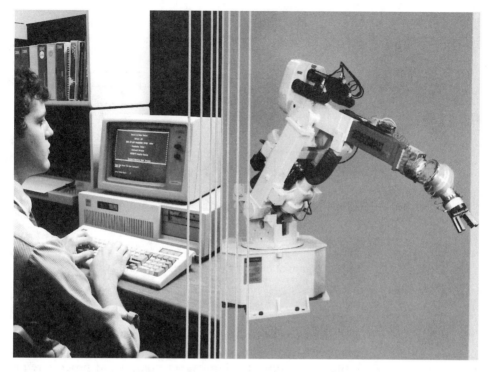

Figure 9.16 The PC controls a robotic system. (Courtesy Cincinnati Milacron)

9.6 Distributed Control

The distributed control system (DCS) involves two or more—usually several—computers in different locations. The functions of a distributed control system may include data acquisition, digital computers, and PLCs that offer digital control; and dedicated controllers offering real-time control. Safety interlocks will also be present. A typical model for distributed control is diagrammed in Figure 9.17.

The blocks in Figure 9.17 are as follows:

- The **data acquisition system** records system data at regular intervals; this task would normally be handled by a PC. The operator may enter programs for managing data or accessing process data.
- The **digital controller,** probably a programmable logic controller located on the factory floor, offers real-time control for sequencing of operations. Digital computers require encoding and decoding, known as analog-to-digital conversion (A/D) and digital-to-analog conversion (D/A) respectively.

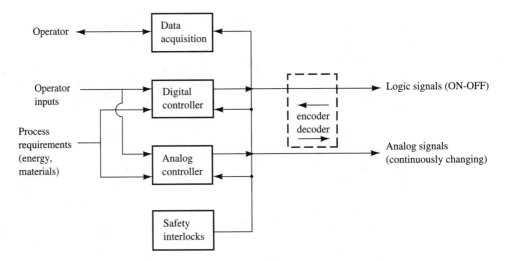

Figure 9.17 A block diagram of one type of distributed control system.

- The **analog (continuously variable) controller** is like the home heating example—quite different from the digital controller. Both types of controllers have operator inputs (set points) and process requirements measured by sensors.
- The **safety system** usually is made up of mechanical relays and switches that are highly reliable and easily inspected.

Troubleshooting

This chapter has shown the computer to be a powerful industrial tool, used by technologists to make decisions and solve problems that could be solved in other ways, given enough time. The computer is fast and controls large amounts of data, but it cannot think or solve problems it was not taught (programmed) to solve. It must also receive correct data to arrive at correct solutions. "Garbage in—garbage out" (GIGO) is the description computer personnel used for invalid data inputs and the sometimes disastrous results.

The technician working with today's industrial computer is responsible for the data inputs—whether the inputs are from process sensors requiring knowledgeable and skilled maintenance or are operator inputs from the keyboard or light pen. When "garbage" does enter the process, the technician applies troubleshooting procedures to correct the problem.

With experience, the technician develops skills and knowledge in troubleshooting. The knowledge is acquired by a combination of practice in solving problems and repairing malfunctions, and a thorough knowledge of how the malfunctioning device is supposed to work.

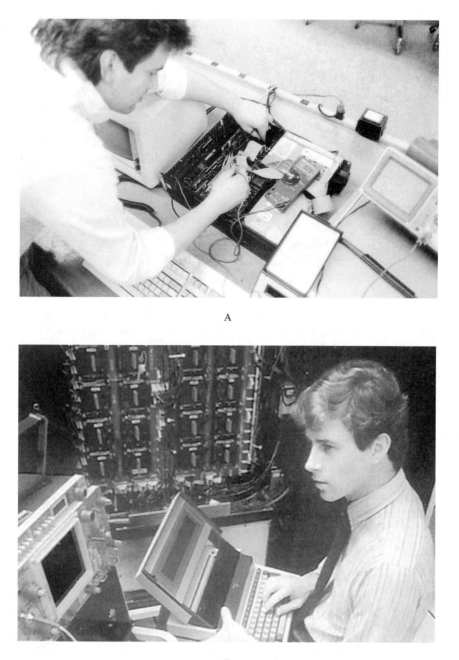

A

B

Figure 9.18 Technicians in industry may use sophisticated electronic measurement equipment and even other computers with expert systems in troubleshooting and repair of PCs and their equipment interfaces. (A, courtesy Central Ohio Technical College; B, courtesy Honeywell, Inc.)

Tools used in troubleshooting include equipment documentation, diagnostic software, and modems that can access remote diagnostic service units. Service technicians will, in addition, definitely wish to acquire social skills to enhance customer relationships.

Please realize that you need not fear computers or their effect on everyday life. Remember that the computer is a tool, like any other tool. The skill of the tool user determines the effectiveness of the tool and usefulness of the end product. A quality technical education provides a sound basis for acquiring the skills and knowledge to maintain and control computers (Figure 9.18).

Problems

Section 9.1

1. Explain what the computer message "syntax error" means.

2. How many unique states can a 16-bit microprocessor have?

3. How many unique pieces of information can be transferred at one time by a 32-bit data bus?

4. Compare and contrast a machine language and an assembly language. Why are neither of these languages considered a high-level language like BASIC?

5. Describe the computer you are using. Is it a PC? If so, what brand and model number? What type of operating system does it have—and what version? Does it have a built-in interpreter? Does it support compiled BASIC?

Convert the BASIC expressions in Problems 6–12 to conventional algebraic expressions and evaluate the expressions with your calculator:

6. PRINT 4E3 + 8E4 − 5E3

    Example: $4 \times 10^3 + 8 \times 10^4 - 5 \times 10^3 = 7.9 \times 10^4$

7. ? 2+6/3−2

8. ? (2+6)/(3−1)

9. ? 15.6E3 * 12E6 * 32E1

10. ? 12E−4 * 1.6E−8 ÷ 2.77E12

11. ? SQR (32 ^ 2 + 18 ^ 2)

12. ? 16 ^ 3/2

Section 9.2

13. If $A = 3, B = 4$, and $C = 5$; what is D$ equal to if D$ = A*B*C?

14. Write a BASIC program to find the area of a rectangle for sides L and W. Specify the input of L and W to be in meters.

15. What will be the output of the following program when run?

    10 PRINT "Your name"

20 GOTO10

What is the result of a loop?

After running this program, try adding the lines:

5 For I = 1 to 40

25 Next I (you have added a FOR-NEXT loop)

The added lines will help you avoid a troublesome, never-ending run sequence.

Write a BASIC program to solve for the torque on a bolt. The equation is $T = CDP$, where T = torque in inch-pounds, C = a constant friction factor of 0.30, D = the diameter in inches, and P = the bolt load in pounds. Solve for the torque on the bolts in Problems 16–18.

16. $D = 0.25, P = 1400$

17. $D = .5, P = 5500$

18. $D = 1.5, P = 75\,000$

*19. Add to the program for Problems 16–18 a test to determine if the bolts will break. The equation for breaking load in pounds is $P = 65\,000 \frac{\pi}{4}d^2$. Which of the bolts will break?

Solve the trigonometric functions in Problems 20–22 with BASIC (remember to convert to radian measure first by multiplying by π rad/180 deg).

*20. Sin 30° =

*21. Cos 30° =

*22. Tan 45° =

Using BASIC, solve the arc functions in Problems 23–25 for their values in degree measure (use 180 deg/π rad to convert the answer from radians to degrees—see Appendix D).

*23. $\sin^{-1}(0.524) =$

*24. $\cos^{-1}(0.785) =$

*25. $\tan^{-1}(1) =$

26. Explain in one paragraph how a compiler is able to speed up the run time of a program.

27. Refer to the discussion on QBASIC and rewrite the BASIC program to convert vectors. Use the program to convert the following vector from rectangular to polar: $X = 4, Y = 3$. Use the program to convert the following vector from polar to rectangular: $5 \angle 45°$.

28. Compile the program in Problem 27. Run the compiled program from DOS. Convert the same vectors as before and note whether or not there is an increase in run-speed for the compiled version.

---

*Challenging problem.

29. Select one of the computer languages in Appendix E. Locate a reference source on the selected language in your college library, and write (or copy from the source) a few program lines. Compare the line commands with similar commands in BASIC.

Example for PASCAL:	writeln ('the average =', average);
in BASIC is written as:	PRINT "the average ="; average

Section 9.3

30. Draw a control model of the heating system (see Figure 9.7) in your home, place of work, or college. Identify the exact type of sensor in the feedback loop (e.g., bimetallic, mercury switch, or other).

31. What is a *set point* in a process-control loop?

32. What does CNC stand for? How does a CNC machine differ from an NC machine?

33. Locate a magazine article addressing process control by using the *Applied Science and Technology Index.* Summarize the article or provide a copy to your instructor.

34. Use Figure 9.9 to write specific activities that might be taking place at each workstation location. Use an automobile company as an example. For instance, in the design engineering position, "the radiator assembly is being designed for a new model car."

35. List the elements leading to CIM systems as shown in Figure 9.6. Do not use acronyms, but write out each type of system. Provide one example of each element (may require library research).

Section 9.4

36. What does PLC stand for? What special logic does the PLC use that electrical maintenance personnel are familiar with?

37. List at least three disadvantages of using mechanical relays to "make" or "break" electrical circuits.

38. Be able to describe the "start action" and "stop action" of the ladder diagram in Figure 9.14 at any time when examined by the instructor.

39. What happens when the start pushbutton is pressed in the ladder diagram of Figure 9.14, and the bottom set of CR1 contacts are normally closed (N.C.)?

Section 9.5

40. List at least three uses for the personal computer (PC) in industry.

41. Discuss, in class, the advantages and disadvantages of PCs compared with PLCs in industry.

42. Using the *Applied Science and Technology Index,* find one magazine article on distributed control and submit it to your instructor.

43. What kind of instrument does the technician in Figure 9.18 appear to be reading? Why would a technician, troubleshooting and repairing a complete computer system (an entire workstation), require mechanical as well as electronic skills and knowledge?

Selected Readings

Adamson, Thomas A. 2nd Edition *Structured BASIC Applied to Technology.* 2nd edition. Upper Saddle River, N.J.: Prentice Hall, 1993.

Murphy, Pat. "PCs and PLC's: Partners in Control." *InTech* (July 1988): 33.

User's manuals accompanying your version of BASIC.

Webb, John and Ronald Reis *Programmable Controllers.* 3rd edition. Upper Saddle River, N.J.: Prentice Hall, 1995.

10

Your Future in Technology

Your future contributions in technology will be vital to our society. The world is changing rapidly, and technology is the engine for that change. A nation lacking leading technology will be dependent on others that are in the lead. Rapid change must also be understood, or the technology may control us rather than we controlling the technology. In order to contribute most efficiently to our technology you must be ready for the challenge of learning a seemingly insurmountable amount of information, often based on a good understanding of the math-sciences. Juanita's experiences that follow may help you to hang in there when the going gets rough.

At the end of her first quarter of study, Juanita came to her advisor's office to officially withdraw from the electronics engineering technology program. The advisor—who was also her instructor in direct current (DC) circuits—was surprised, because he knew she was handling the circuits material well. After the instructor asked about her grades in other courses Juanita assured him that grades were not the problem, even though she was a single parent working part-time who had been out of school for ten years. Under these stressful conditions, the B average she expected represented outstanding performance.

"I feel it's impossible to be a successful tech," the student exclaimed. "There is just too much to learn and remember. It is impossible to know it all; and, in a man's world, it will be even more difficult to have the self-confidence I need to survive even the first year."

That was three years ago. Today, Juanita has survived her first year in industry and has received an early promotion.

Her college instructor and academic advisor explains what took place that day, three years ago.

Juanita was correct, *according to her assumptions at least,* of what a technician's job entails. She was fearful and overburdened with information, feeling she could never achieve the goal of becoming a qualified technician.

Were Juanita's assumptions correct? They were not. Most instructors have witnessed this fear of failure in many aspiring technicians and technologists during the first year of study. This fear occurs with males as well as females, with students just out of high school, and with the returnees who come later with "rusty skills" rather than "no skills." Many times it comes as a question in the laboratory: "If I can't understand this simple circuit—only three resistors and a power supply—how can I ever understand an infinitely more complex computer system?"

But the real key to solving Juanita's dilemma was to explain to her the need for specialization. Juanita learned that her education and training would not end after graduation. She would be trained on specific equipment and/or processes in her new position. The basics she had learned would be vital to her understanding, but she would not need to recall everything and could look much up in former college texts and technical handbooks.

Juanita listened that day and made the right decision to stick it out. Her computer programming course in the spring quarter provided the focus she was searching for.

After her first full year on the job Juanita still has to look up previously learned circuit fundamentals when interfacing PCs to control industrial processes. But her supervisor already feels she is "one of the best networking technologists in the plant."

Your main purpose in reading this text, and completing the associated course work, has been to gain a better understanding of the engineering technologist's place in this dynamic world. Many of the tools used in a technical career have been discussed, including calculators, computers, and the math-sciences they are based upon. Your reading thus far demonstrates your enthusiasm for learning these powerful tools and a commitment to pursue a technical career.

The challenge now is to choose one of the engineering technologies listed in Chapter 2 or even a more specific technology existing in your area. This initial step—narrowing your career choices—is only a beginning. Many successful technicians become quite specialized—extremely knowledgeable in one specific area. Even one narrow, well-defined area in the technologies can be extremely challenging.

Examples of a few of these more specialized areas are

- programming computers to automate processes;
- maintaining numerical control centers;
- assisting in superconductor research;
- designing robotic applications;
- installing machine vision systems;
- determining the appropriate blend of chemicals in a batch process, resulting in a composite material with improved properties, such as higher strength and lighter weight;
- testing the accuracy, reliability, and safety of implanted medical devices (e.g., prostheses);
- calibrating laser power outputs; and
- installing fiber-optic systems.

The future will be quite challenging, profitable, and rewarding for those technicians or technologists who become intimately involved with their specialty area. Technicians are needed to implement new technologies so society may progress. Wanting to contribute to a better life for others is a goal we all share.

Futurists, those who specialize in predicting the future, often use **content analysis** to identify the current forces that will effect change for the future. Trends in the amount of media coverage devoted to certain topics give keys to the direction the world, a country, or a particular technology is headed. The need for you to try to anticipate the future and future careers cannot be overemphasized.

The following sections represent a survey of some emerging technical areas offering great potential for the future. Some of these areas are still in flux and in the hands of the research scientists. Soon they will be in your hands—the hands of the implementor.

The selected technologies are by no means an exhaustive listing, but familiarity with these areas may help you to focus your future in the engineering technologies. The topic areas are the computer, robotics, optical systems, composite materials, and opportunities for service in areas of world need.

10.1 The Evolving Computer—A Base for High Technology

To the technician and technologist the advance of the computer is the basis of most of what is termed high technology. Robotics, computer-aided design and drafting (CAD), computer-aided manufacturing (CAM), and other high-tech systems will become as advanced and extensive as the computers and software controlling their operations allow.

Advances in computer technology seem limitless. Faster computers with more memory continuously replace smaller, slower machines in the marketplace. Growth today leads to even faster growth in the future as new technology becomes available (Figure 10.1).

Changes in the **personal computer (PC)** arena within the next 10 years may include

- faster processors at lower cost (processor speeds will approach 500 MHz)
- smaller PCs (expect pocket-sized mobile devices)

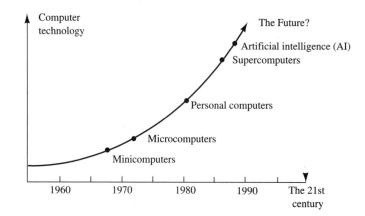

Figure 10.1 More technology leads to faster growth of technology; an exponential growth in computer technology results.

- **virtual reality systems,** allowing users fitted with goggles, helmets, gloves, and other tools to project bodily motions into the world of the computer.
- more network computers (NCs) designed to use the power of network servers (Figure 9.9) and the internet
- more imaginative software that will be easier to use

Supercomputers represent the state of the art in fourth-generation computers. Some supercomputers have up to 16 background processors and cost over 15 million dollars. Even at these prices customers are waiting in line to purchase the powerful machines. Figure 10.2 is a photograph of the Cray-2, one of the world's most advanced supercomputer systems.

Figure 10.2 This Cray-2 is in operation at LLNL's National Magnetic Fusion Energy Computer Center in California. (Courtesy Lawrence Livermore National Laboratory)

The federal government has invested heavily in supercomputer research, mostly in applications to support aerospace technology. The F-15 aircraft carries a computer occupying less than one cubic foot of space, yet it can process more information than 200 000 personal computers. Sandia National Laboratories in the United States has a series of computers, called SANDACs, that may be as powerful as early Cray computers but are much smaller and much less expensive. SANDACs are now as small as an automobile battery and may be built for less than $500 000.

Artificial intelligence (AI) is a new discipline still developing. New supercomputers designed for AI capabilities represent a top priority for many nations, most notably the United States and Japan. A "thinking machine," the **fifth-generation computer,** may require computer memory cells to be sized down to the molecular level. Cryogenics, or super-cold technology, will lower the temperature of computer elements to preclude meltdown.

The AI supercomputers will take a cybernetic approach to information processing. The science of **cybernetics** applies the knowledge we have of an animal's nervous system to automated control systems. One important outcome of cybernetic research in the late 1940s was the feedback loop discussed in Chapter 9. The cybernetic approach will require new ideas in information processing and computer language structures.

One currently useful application derived from AI technology can be processed on smaller computers—even PCs. The **expert system** collects diagnostic information from experts in a particular field and programs a computer so an application user can query the program and diagnose a complicated system (Figure 10.3). For example, experienced physicians, via the expert program, can lead a patient to a successful diagnosis of an ailment by simply responding to questions about their perceived symptoms.

On a more practical level, technical positions are increasing in the **office environment.** Rapid growth in computer technology has led to increased automation and computerization of the office. Office computers now read, copy from, and communicate with other computers (Figure 10.4).

To appreciate the rapid growth in office technology, take, for example, one type of office machine. **Facsimile (fax) machines** convert images of the written word as well as maps, pictures, and graphs into digital signals to send over telephone systems to other faxes, which reconstruct the image. Fax machines are available now for one-fifth the cost of similar models five years ago. Of course, as costs come down for office equipment, more technology is put into place.

Technicians are required to service the growing numbers of computer-based equipment. Without quality technical support, office productivity will be seriously jeopardized.

A computer-based technology that may rival the fast-growing cellular phone market is called the **global positioning system (GPS).** GPS is based on a constellation of 21 high-altitude orbiting satellites. The satellites are like artificial stars, which replace the stars used previously for navigation (Figure 10.5).

The GPS works by timing how long it takes for a satellite signal to reach a receiver (satellites have atomic clocks on board). Once the time to reception is

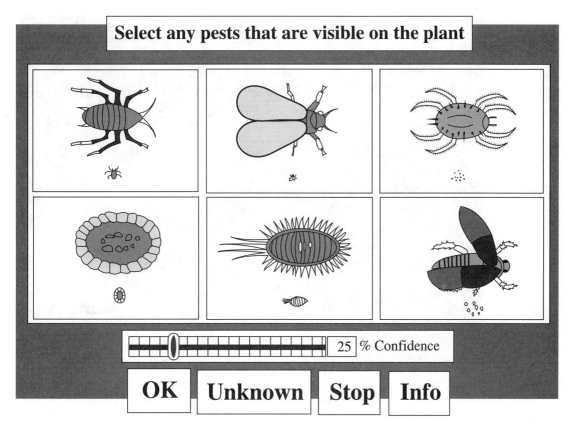

Figure 10.3 An expert system is used to identify insect pests.

known, the distance from the satellite can be determined by dividing the velocity of light (186 000 miles/second or 3×10^8 meters/second) by the time it takes for the signal to reach our receiver. In Figure 10.5A, one satellite signal received can give us distance information so that we know that we are within, say, 11 000 miles of that satellite. Figure 10.5B shows that distance information from two satellites fixes us (our receiver) in a much smaller circle. Figure 10.5C, showing triangulation by three satellite signals, puts us at one of two points. It turns out that one of the points will be eliminated because it is ridiculous—too far above the earth, in the ocean for a land vehicle, etc. (Four satellites are used in actual practice, and a receiver on a very fast jet aircraft may have four channels to obtain an instantaneous fix on its location.)

GPS receivers, like cellular phones, will become cheaper and more common. Applications include providing emergency and delivery vehicles (eventually even personal automobiles) with electronic (geographical information system [GIS]) maps that will show the way to any destination; preventing air and sea collisions

Figure 10.4 Office computers perform such tasks as business planning, interoffice communications, and the electronic distribution of documents. (Courtesy International Business Machines Corporation. Unauthorized use not permitted.)

by precise avoidance systems; and making zero-visibility aircraft landing systems feasible. The latest Differential GPS (DGPS) systems use two receivers—one stationary at a known location and the other at an unknown location. DGPS is so accurate it may be used to track movements in the earth's crust along an earthquake fault.

The computer age is here and most computers are invisible, in the form of hidden microprocessors connecting us to our home appliances, office machines, and the automobile we travel in. All of these chips will eventually be connected through the information highway, so we can control devices from afar and communicate with each other more efficiently. The **information highway** is a communication system composed of

- the telephone companies' infrastructure,
- the cable companies' infrastructure,
- and the internet.

Most of the highway is already reality in the form of existing telephone and cable companies (Figure 10.6A). What remains is to sort out which companies will provide

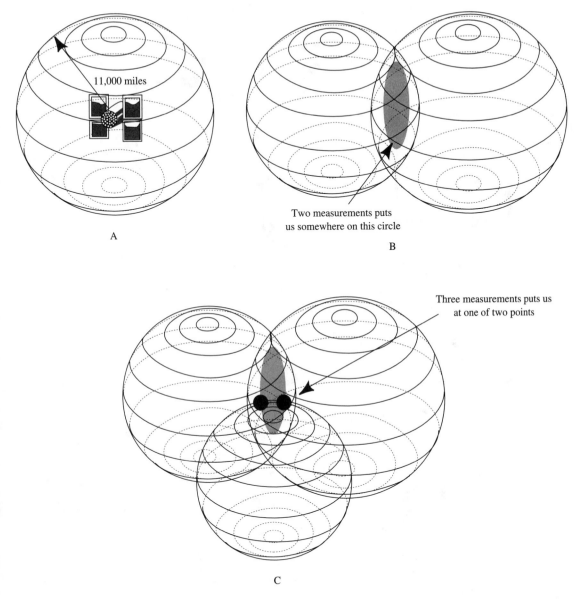

11,000 miles

Two measurements puts
us somewhere on this circle

A

B

Three measurements puts us
at one of two points

C

Figure 10.5 GPS uses at least three satellites to triangulate a position. The spheres represent the distance from the satellites, not the earth's surface.

what services. Service providers must be able to charge in order to make their investments economically sound. Figure 10.6B shows what the future information highway might look like.

The extensive developments in telecommunications may include, but be not limited to

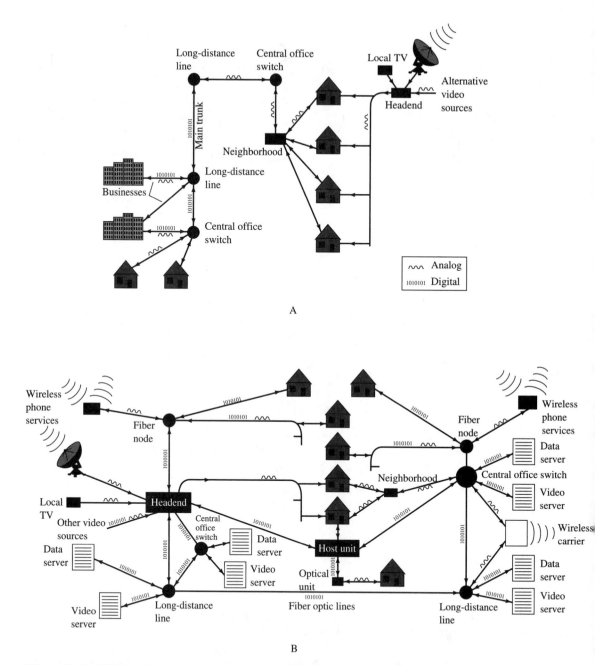

Figure 10.6 (A) Today's telephone and cable systems. (B) Tomorrow's information highway. The servers are supercomputers with hard disks that customers may access. (A, B: Adapted from original figures. BYTE Magazine, pp. 48 and 49, March 1994 © by McGraw Hill, Inc., New York. All rights reserved.)

Figure 10.7   A 16-bit sound card digitizes analog sounds. (Courtesy International Business Machines Corporation. Unauthorized use not permitted.)

- **further digitization,** or the conversion of words, numbers, sounds (Figure 10.7), and visual images to binary. For instance, **personal communications services (PCS)** will replace the analog (continuous-wave) cellular phones with digital cellular phones. A digital signal is clearer than an analog signal.
- **friendlier input devices,** which will mean the elimination of keyboards and mice that require knowledgeable users and impede timely access to information.
- **increased mobility,** in the sense of smaller computers and embedded microprocessors in other equipment and appliances. These embedded systems may someday communicate through the internet, and, while on vacation, you may be able to program your VCR from the Net.
- **improved networking systems,** including faster modems and recent advances in technologies such as an **asymmetric digital subscriber line (ADSL),** or an **integrated services digital network (ISDN).** Both technologies offer ultra-speedy phone lines that will improve downloading time. Other networking options are cable systems that can handle two-way communications and satellite services that can beam the Net down to you from space.

The various companies on the information highway will have to accept standardization so that the devices that link people and equipment can talk to each other. Most of this standardization has already been worked out by an international computer cooperative known as the internet (Section 8.4). The **internet** is a set of protocols or codes that allow computer users access to vast amounts of diverse knowledge provided by universities, professional groups, and government agencies. The basic research that has resulted in the protocols has been funded by the U.S. government; the most recent legislation is the National Research and Education Network Act sponsored in Congress in 1991 by then Senator Al Gore. You may access the internet today by using a personal computer and modem. But the Internet will be used in the future by millions of customers that have never used a personal computer and may never wish to.

The highway represents a very large investment by the U.S. and other industrialized nations. Business is willing to invest in order to enhance communications with customers. Better communications with customers increases the range of data needed to better understand customer needs and the way the marketplace is meeting those needs. It will also allow better communications with suppliers, government regulators, financial institutions, and so on. Government will encourage investment in the information highway because it will stimulate economic growth. In addition, good communications between government officials and the people they serve are central to the effectiveness of democratic governments.

Someday, customers may use their TV sets to request a video of their choice to be played at their desired time or to use a multimedia encyclopedia. They may use the same TV, or even palm-sized personal communicators, to access current information on their bank account, the stock market, or local traffic conditions. They may even be able to vote by turning to the appropriate channel.

The huge increases in investment, billions of dollars, will ensure the need for many implementors. Technicians and technologists will be needed to install new systems and maintain those in place. Maintenance activities will include the hot-swapping of failed components such as hard drives and circuit boards, along with the upkeep of the many high-speed servers that will connect customers to audio, video, and data services.

10.2 Robotics

To most of us, the terms *robot* and *high technology* are synonymous. The need to automate has been discussed (Section 1.3). The robot will play a large role in automating manufacturing as we enter the next century. The Robotic Industries Association (RIA) reported $158 million in value of product shipped from January through March of 1994. This figure was a 60 percent increase over a comparable period in 1993.

The challenge to the technician is to help decide whether to install a robot in a particular operation (Section 3.3), aid in selecting the right hardware and software if the conversion is feasible, determine how it is to be installed, and develop a maintenance plan for the system.

The robot itself most often consists of a mobile arm with finger-like appendages. It is controlled by a microcomputer that may be reprogrammed anytime. The RIA first defined the robot rigorously as:

A reprogrammable, multifunctional manipulator designed to move material, parts, tools, or specialized devices, through variable programmed motions for the performance of a variety of tasks.

The technician may best understand a robot by considering its parts and their specifications or applications. The gripper, **end effector,** or end-of-arm tooling is a

Table 10.1 Typical Applications of Robot End Effectors

Type of Effector	Typical Application
Gripper	Pick and place
Vacuum or magnetic pickup	Handle delicate parts
Welder	Spot/continuous weld
Sprayer	Paint spraying
Torch	Cutting

Table 10.2 Typical Controller Applications

Point-to-Point	Continuous-Path
Parts transfer	Spraying operation
Gluing and sealing	Polishing and grinding
Loading and unloading	Precision assembly
Spot welding	Arc or spot welding

good place to begin (Table 10.1). Also discussed below are other major robotic components: the manipulator, or arm; the controller (Table 10.2); and the power supply.

One way robots are classified is by the number of motions the **manipulator** can make. Both rotational motions about an axis and linear motions along an axis are counted. The number of independent ways a robot can move is also known as the robot's **degrees of freedom** (Figure 10.8). (Can you determine the number of degrees of freedom for the robotic system in Figure 3.10A?)

One southwestern robotic manufacturer has specialized in less sophisticated, non-servo robots (Figures 10.9A and 10.9B). **Servo-controlled robots** use continuous feedback. A **non-servo robot** has only switches or physical stops at end-of-travel. When the arm makes one move it cannot move again until the limit-stop is reached or a position switch is activated.

Non-servo controlled robots are inexpensive and versatile. Most are used for **pick-and-place** operations. Additional advantages are that they have high accuracy and reliability, are easy to understand and program, and are simple to maintain. Disadvantages, when compared to servo-controlled robots, are that they have a limited number of stop positions and complex trajectories are not practical.

Another way robots are classified are by their work envelopes. Typical work envelopes take the shape of cylinders (three degrees of freedom) to spheres (six or more degrees of freedom).

The **controller** acts as the brain of the robot. Just as the human brain causes motions of arms and legs, the robot controller causes manipulator motion. Depending on the sophistication of the controller, manipulator motion may be point-to-point (non-servo) or a smoother moving continuous path (servo-controlled).

The controller (Figure 10.10A) is often a mini- or microcomputer. The software for the computer controller enables the operator to "teach" the robot what movements to make. The operator moves the robot arm, or a lighter "teach wand," through a series of motions that will accomplish the given task. The controller records these motions and, once taught, can repeat the task indefinitely.

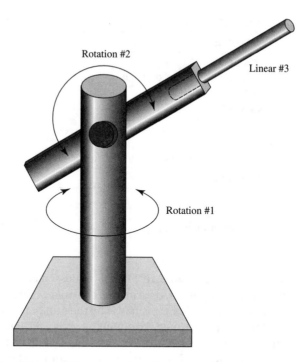

Figure 10.8 The degrees of freedom of a robot manipulator or arm include the rotational and linear movements of the arm.

The **power supply** may be likened to the human body's muscular system. Power transfer for a robot is accomplished by pneumatic systems, electric motors, or hydraulic systems. Many robot designs include a combination of these power-transfer systems.

The **pneumatic system** is generally used for low-cost, less-accurate pick-and-place operations. This is due to the nature of pneumatics. Hard to control in a linear fashion, it is used mainly in limit-stop to limit-stop operations. A complete pneumatic system includes a filter, regulator, lubricator, and air compressor.

Electric motor drives are used where greater accuracy is preferred (Figure 10.10B). The electrical drives are more compatible with the digital electronic controllers discussed in Chapter 9. However, loads cannot exceed several hundred pounds.

For very large loads, **hydraulic power supplies** are used. Hydraulic systems are expensive because they include pumps, reservoirs, and heat exchangers. The advantages of hydraulics include the ability to handle large loads and precise control.

You may become more sensitive to the difficulty of designing a functional robotic system by role-playing a robot. Begin by smearing petroleum jelly on your glasses (safety glasses will do), tie one hand behind your back, place a mitten on your free hand, and proceed to assemble a snap-together plastic model (example from Joe Engleberger, the father of robotics).

The technician and technologist should be able to list the major components of a robot and have some idea of the applications found in the tables and discussed

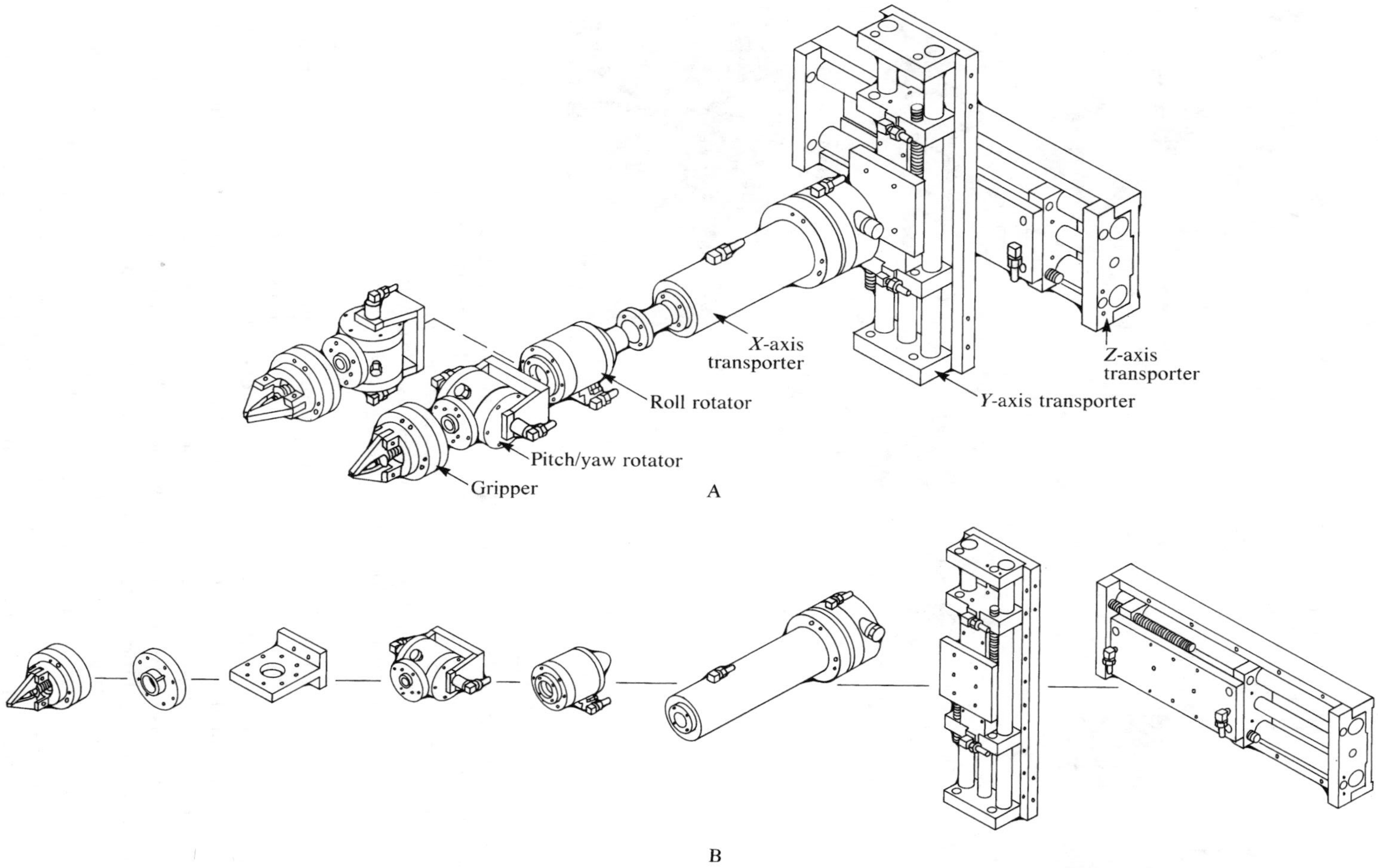
X-axis transporter
Z-axis transporter
Roll rotator
Y-axis transporter
Pitch/yaw rotator
Gripper
A
B

Figure 10.9 Simplicity is often the key to the successful implementation of robotics in industry. (A) A non-servo robot, that (B) uses a concatenated approach (may be built up of selected serial components). (Compliments Mack Corporation, Flagstaff, AZ)

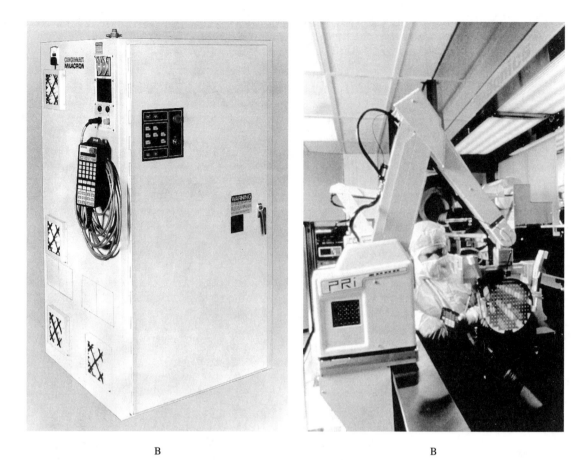

B   B

Figure 10.10 (A) This programmable robot controller is housed in an industrial cabinet, environmentally controlled and dust-free. The teach pendant attached to the side of the cabinet is used to "teach" the robot to follow an operator-directed controlled path. (Courtesy Cincinnati Milacron) (B) Precision electric motors drive the robot used in chip manufacturing. (Courtesy International Business Machines Corporation. Unauthorized use not permitted.)

above. It is important to realize the robot performs only 20 to 30 percent of the manufacturing process. To simply install a robot where a production worker once stood and activate a few controller commands is not enough. The function the robotic system is intended to perform, its planned maintenance, fixturing and tooling, and—above all—safety precautions must be carefully considered before installation.

The world of robotics is still new and exciting. Thirty years ago computers filled huge rooms, had little processing power, and needed constant maintenance. Today's computers are infinitely more reliable and are used everywhere. Robotics will follow the same path. For instance, one Swiss manufacturer makes a robot that can crawl through sewer lines and repair them. Costs savings are tremendous because no excavation is necessary. The robot can smooth and clean the sewer lines and repair them

Figure 10.11   A robot known as the "mole" is used to check on the presence of a burrowing owl. (Courtesy Lawrence Livermore National Laboratory)

with epoxy cements, while providing the operator with a video of the work. Today's robot manufacturers are focusing on such specialized roles for robots as helping hospital patients move around and extend their reach, exploring the ocean floor, handling hazardous waste or defusing bombs, or even to checking out the burrows of ground-dwelling animals on the endangered species list (Figure 10.11). Someday, with technicians and technologists available who can install and maintain reliable systems, the robot will be just as indispensable as the computer is today.

10.3 Optical Systems

How would you like to hitch your future to a star? A star's light, that is.

One of the newest, most rapidly growing technologies, offering great force for future change, is optical technology. **Optics** is the study of light—invisible as well as visible light. Studies include how light is produced, transmitted, and measured and how it is converted to other forms of energy. Advances in optical systems are being made in

1. electrical generation with photovoltaics,
2. communication systems using lasers and fiber optics, and
3. optical computing systems.

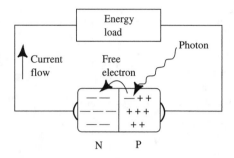

Figure 10.12 The semiconductor forms the basis of one photocell. The N-type region has excess electrons and the P-type an absence of electrons.

Photovoltaics

The photovoltaic (PV) industry may someday rival the size of other electrical producing industries: nuclear fission, hydroelectric power, and fossil fuel steam-generating plants. Unlike other energy sources, PVs produce clean, safe, low-maintenance electricity.

Albert Einstein won the Nobel Prize for physics in 1921 for his theories on the photoelectric effect. He showed that "quanta" of light—small energy packets now called **photons**—striking atoms in a metal can cause electrons to be released. A **photovoltaic cell** enables the photoelectric effect to convert the sun's energy into electricity.

As shown in Figure 10.12 the light energy-packed photon frees charges to flow across the charge barrier between the N- and P-type materials. Connecting a load to the semiconductor will permit charge carriers to flow. Efficiencies are low today, but higher efficiencies are expected. As an example, current PV systems, installed, cost $5 to $7 per watt produced. But in the early 1970s, PV electricity production cost $100 per watt. Industry trends and exciting new developments lead technologists to believe in the future of this technology.

The manufacturing of PV cells, a high-technology industry, is expected to increase. The technology will depend on our ability to form high-quality (impurity-free) crystals as continuous thin sheets or ribbons at high rates.

The present goal is to produce photovoltaics that yield up to 20 percent efficiency. **Point-contact PV cells,** a relatively new technology, could increase efficiencies up to 25 percent. The theoretical limit for silicon-cell efficiency is around 32 percent.

Many technical colleges are leading the way in applying photovoltaic technology to real-world situations. For example, Stark State College of Technology, an Ohio technical college, built a solar car that was selected to compete in the prestigious Sunrayce of 1990 (Figure 10.13). Sunrayce '90, the longest and largest solar car race ever held, was a grueling 1800-mile, 11-day race over U.S. roads and highways from Orlando, Florida to Warren, Michigan. The drivers, four men and one woman, tolerated 95- to 120-degree temperatures and participated in a physical training program before the race. The high-quality solar cells used were 13.5 percent efficient. Despite competition from colleges and universities with larger solar car budgets, the team placed 13th out of a field of 32. In Sunrayce '93 solar cell efficiencies were

Figure 10.13 A solar car designed and built by students of Stark State College of Technology. (Courtesy Stark State College of Technology)

already improved to 17 percent, demonstrating the rapid advances being made in the technology.

Photovoltaic industry spinoffs you may choose for a career include *electrical conversion systems* to convert the PV cell's DC power to AC power, *more efficient batteries* to store the DC power, and *more efficient marketing and distribution strategies* that will drop prices to the point where homeowners can afford to install photovoltaic systems.

The Laser and Fiber Optics

The acronym **LASER** stands for *L*ight *A*mplification by *S*timulated *E*mission of *R*adiation. There are many technical applications for laser technology. Examples of the varied applications include precision measurement systems, communication systems, vision systems, fusion research, welding, cutting, performing delicate surgery, and reading bar-code information.

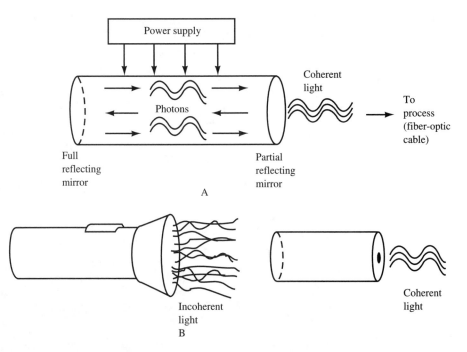

Figure 10.14 (A) A power supply pumps up electrons to produce photons of light energy inside a laser tube. (B) The incoherent, diffused light of a flashlight is contrasted with the coherent, concentrated light of a laser.

Refer to Figure 10.14 to understand how the laser works. The power supply creates an electromagnetic field around the laser tube. The force field created is said to "pump-up" the electrons in the laser material to high energy levels. As the electrons return to lower energy levels, they emit photons of light energy.

The photons rebound from mirrors at each end of the laser tube. The reflections are synchronous with the power supply's frequency and are amplified within the tube. Coherent light results. **Coherent light** waves are the same frequency, producing monochromatic (single-color) light, and are in phase with one another.

The partially silvered mirror at one end of the laser tube allows some of the coherent light to escape. This coherent light at a single frequency is extremely intense and tends not to diverge (it forms a narrow beam). Lasers are classified by the materials used to make up the laser tube. Table 10.3 lists some common types of lasers, with the year of development and a typical use for each.

Recent research has resulted in **upconversion lasers.** The new lasers take advantage of some elements to absorb and share energy under certain circumstances so that some of them reach the very high energy levels needed to produce blue light. Blue light has more than twice the frequency of the infrared light used to initiate laser action. The higher frequency means greater data-handling ability. Semiconductor diode lasers, the most common and most efficient lasers, may someday be able to pump blue light upconversion lasers (Figure 10.15).

Table 10.3 Types of Lasers and Typical Uses

Helium-neon (1962)	First commercial laser; used in communications and measurement systems
Ruby (1963)	A crystal laser used to melt hard materials
Neodymium: glass (1964)	Emits invisible light; used in experiments with plasmas (e.g., nuclear fusion)
Neodymium: yttrium-aluminum-garnet (Nd : YAg) (1964)	A crystal laser used as a drill or rangefinder
CO_2 (1965)	A powerful gas laser used in military applications
Semiconductor-diode (1970s)	Operating at room temperature, these "laser diodes" offer low-cost products used in compact disc (CD) players and in fiber-optic technology. Light-emitting diodes (LEDs) are also used in fiber-optic communication systems.

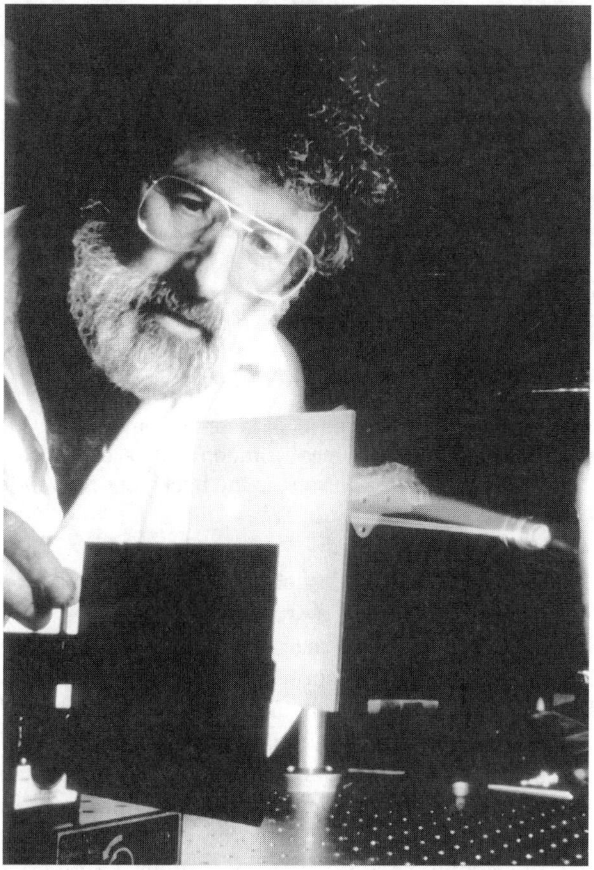

Figure 10.15 The development of the upconversion laser demonstrates the ongoing research in laser technology. (Courtesy International Business Machines Corporation. Unauthorized use not permitted.)

Figure 10.16   The beams from the huge NOVA laser system (over 600 ft long) will converge simultaneously onto a small heavy-hydrogen fuel pellet and create a fusion reaction. A laser technician can barely be seen in the background. (Courtesy Lawrence Livermore National Laboratory)

Two application areas that bear watching in laser technology are fusion research (Figure 10.16) and vision systems.

Fusion research may bring the world infinitely cheap energy. Fusion fuel is abundant in water. One gallon of water contains one-tenth of a teaspoon of deuterium, which when combined with another heavy isotope of hydrogen (tritium) has the energy of 300 to 700 gallons of gasoline. This means the fusion fuel in the water of San Francisco Bay could supply North America's energy needs for many thousands of years while dropping the bay's water level only an eighth of an inch!

A **vision system** recognizes objects, inspects them, and determines where they are so a machine or robot can access them. Vision system applications include precision measurement and inspection of manufactured parts (Figure 10.17), grading of food and textile products, sorting of products by size and shape, detection of stress regions in glass products, and aerospace guidance. All will depend heavily on laser

Figure 10.17 In addition to being a critical part of a machine vision system, the laser performs precision measurement using laser interferometry. This laboratory technician uses a computer-controlled lathe with laser interferometry to machine parts such as precision mirrors to submicrometer accuracies. (Courtesy Lawrence Livermore National Laboratory)

development. The increasing emphasis on improved quality control makes vision systems one of the fastest growing technologies.

The greatest application area for laser technology will be optical communications systems. A light source can be modulated to carry information, just as a radio wave's frequency can be modulated to carry information from a transmitting station. The modulated light source may be a laser diode or a light-emitting diode (LED).

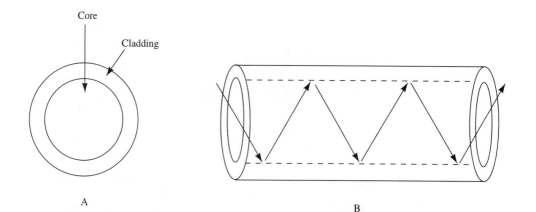

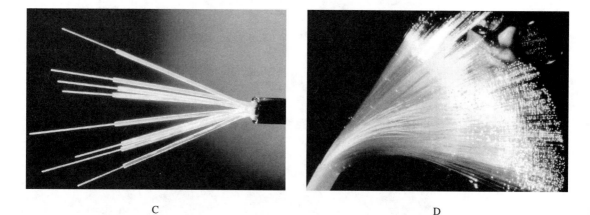

Figure 10.18 Fiber-optic cable. (A) The core is one type of glass, the cladding another. (B) This causes light to be totally reflected as it moves through the cable. (C) and (D) The cable and its materials. (Courtesy International Business Machines Corporation. Unauthorized use not permitted.)

In addition to the laser, the fiber-optic cable is the other major component that makes optical communications possible. The **fiber-optic cable** is manufactured of glass and uses internal reflection principles to conduct photons of light energy (Figure 10.18). Fiber-optic cables were first installed in 1977. Today, millions of miles of fiber-optic cables are in use.

The advantage of fiber optics over electronic transmission is due to the difference between the electron and the photon. Photons travel 50 percent faster through a glass fiber than do electrons passing through a metal conductor. Photons also carry no electrical charge. This means they do not interfere with one another and can even move in different directions at the same time within the cable. See Table 10.4 for a summary of advantages.

Table 10.4   Advantages of Fiber-Optic Transmission

1.   Ease of handling—It is a light, relatively thin material that can be run through the same conduits that now carry copper cable. Its lightness makes fiber-optic cable desirable for avionic systems.

2.   Faster—Optical frequencies carry more information (a single cable may carry up to 10 000 digital television channels at once). Optical energy travels 50 percent faster than electrons passing through copper.

3.   High efficiency—Very low power losses exist in fiber-optic transmission cables. Nonsilica-based fibers may allow unrepeated (unamplified) signals to travel up to 10 000 km.

4.   Isolation—Equipment connected by fiber optics does not require grounding.

5.   Interference is nonexistent—**Electromagnetic pulse (EMP)** interference during nuclear attack and **electromagnetic interference (EMI)** from large motor systems and other EMI transmitters will not interfere with fiber-optic systems.

6.   Reliable—Glass optical fiber is not affected by moisture, temperature, and most acids. Fiber-optic cables continue to send information at up to 2000°F.

7.   Safe—Unlike electrons, photons do not initiate explosions in hazardous areas.

8.   Security—No electromagnetic waves are emitted, reducing the risk of wiretapping and electronic snooping. Fiber optics usually eliminates the need for encrypting military transmissions.

Disadvantages of fiber-optics implementation include connectivity (Figure 10.19), installation problems (bending the cable can break it), the need for special test equipment, inability to transmit analog signals, the need for additional training for technical workers, lack of standards for the technology, and a slight risk of eye damage.

In industry, fiber optics improves communications between computers. These optical communication systems are referred to as **broad-band local-area networks (LANs).** With broad-band LANs, computers can communicate over several hundred feet without losing data transfer reliability. With electronic transmission, EMI can reduce reliable transmission of data to 40 or 50 ft.

Other uses for fiber optics include medical applications, such as optical fiber sensors that can be inserted into the body; fiber-optic transoceanic cables, handling

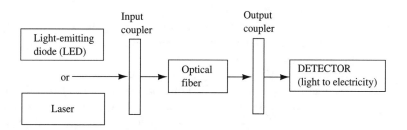

Figure 10.19   A block diagram of the installed fiber-optic system depicts sources (laser or LED), couplers, and final transducer (detector). With experience, technicians can install couplers in about the same time as they do coaxial connectors.

more information with less amplification; and photoelastic transducers (the use of optical fiber material to sense pressure change, measure accelerations, etc.). Many new uses for this exciting technology are being discovered each year.

The challenge in manufacturing fiber-optic cables is to make the fibers clearer and stronger. Already, cables made with silica fiber are so clear and free from impurities they allow signals to be transmitted over 400 miles without repeaters (amplifiers). The high-technology manufacturer who can improve quality and bring down costs will provide many technical jobs for the future.

Optical Computing Systems

One of the latest applications for fiber optics is the "optobuss." The data bus was introduced in Section 9.1. The **optobuss** acts as an optical data bus for high-speed circuitry. An optobuss is fiber-optic material embedded in printed circuit boards. Advantages cited are similar to the advantages for all optical transmission systems: increased computational speed, immunity to electromagnetic pulses, thermal insensitivity, and system reliability. The present speed record for optical computing systems is one trillion bits a second! One trillion bits is equivalent to the information contained in 300 years of a daily newspaper.

The laser is the heart of another optical system—the compact disc (CD) player. The **CD** has brought great stereo music to our homes. Its enhanced sound quality is due to a nonmechanical laser beam reflecting from the surface of the disc, rather than the mechanical contact required of a phonograph needle or magnetic tape head. The result is almost no wear on the optical discs, extending their life almost indefinitely. (Note that the optical disc is spelled with a "c," rather than a "k," as are computer magnetic disks.)

The CD player is only the tip of a new technology iceberg impacting the computer industry. The same technology that put so much quality music on a small disc can also read surprisingly large amounts of data into a computer. For instance, an entire encyclopedia can be written onto one disc. The single-disc encyclopedia contains more than 100 Megabytes (100 MB) of information. **CD-ROM drives** (so called because the discs can only be read and not written to) are available for most popular PC systems (Figure 10.20). These enhanced data bases will become a critical part of computer technology.

Future developments include the ability to write to an optical disc as well as read it. **CD-WORM** (Write-Once-Read-Many, also referred to as CD-R) drives are presently available. Now computer users can create their own CDs. The **digital video disc (DVD)** will eventually replace present day CDs. DVD-ROMs will hold as much as 16 gigabytes (16 GB) of data, compared to the 650-MB capacity of the CD-ROM (about 25 times more storage capacity). The extra storage will allow developers to put a couple of full-length movies on one disc and allow for interaction. For instance, viewers could choose alternate story lines. The development of artificial intelligence (expert systems) will depend on the enhanced memory capability of optical disc technology.

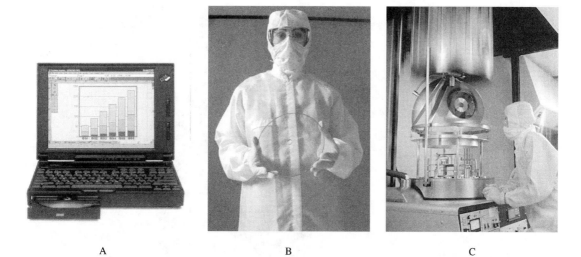

Figure 10.20 (A) The laptop computer sports a CD drive. (Courtesy of International Business Machines Corporation. Unauthorized use not permitted.) (B and C) Making the CD-ROM master requires clean room technology. (Courtesy Metatec® Corporation)

LASER and fiber-optic technology will play an increasingly important role in communication and computer technology. The result is a need for **photonics technicians,** the title given those trained to work with optical systems. The need for photonics technicians will double between 1997 and the end of the decade! All technologists should be aware of photonic applications and be able to list the advantages of optical systems over electrical systems.

10.4 Materials Technology

Perhaps one of the least reported, but most important, high-technology fields is materials technology. Better materials mean improved comfort, safety, and overall quality of life. We have already witnessed the benefits of advanced materials in our automobiles and in our own bodies (e.g., biomaterials—tissues and organs made of ceramics, composites, glass, and plastics).

Some of the high-technology materials that will improve our lives have been around for a long time. Fiber-optic cable is made of the same basic glass material used for centuries in windows, but now fiber-optic glass is so free of impurities that a piece 100 miles thick is clearer than a standard window pane! Concrete can be mixed from new cements so pliable they can be made into springs. Greater flexibility means highways, bridges, and culverts that could last decades longer without maintenance.

Business and industry leaders know how important new materials are. Consequently, research and development (R&D) expenditures for materials research

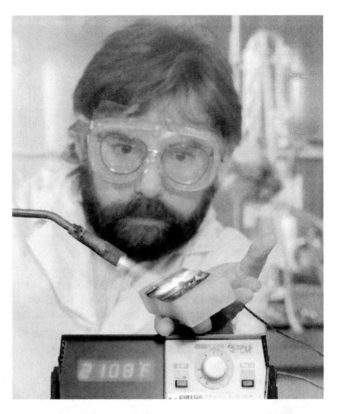

Figure 10.21 One of the best insulators ever tested. A silicon aerogel material is exposed to a temperature of 2108°F! (Courtesy Lawrence Livermore National Laboratory)

are expected to increase more than all other R&D expenditures. Materials R&D includes laboratory activities, such as sample preparation and testing of the material samples (Figure 10.21). Photomicrographs or scanning electron micrographs are critical in materials research (Figure 10.22A).

Industrial activities involve learning how to form and use new materials in manufacturing processes. The technologist will be most important in implementing new processes for the manufacture of advanced materials. The main categories of advanced materials discussed in this section are **high-performance metal alloys (superalloys), ceramics, semiconductors, and composites.**

High-Performance Metal Alloys

High-performance metal alloys are attractive, good conductors of heat and electricity, easily formed, and strong. The technologist responsible for design should know how to research the many different types of alloys and select the one most suitable for each application.

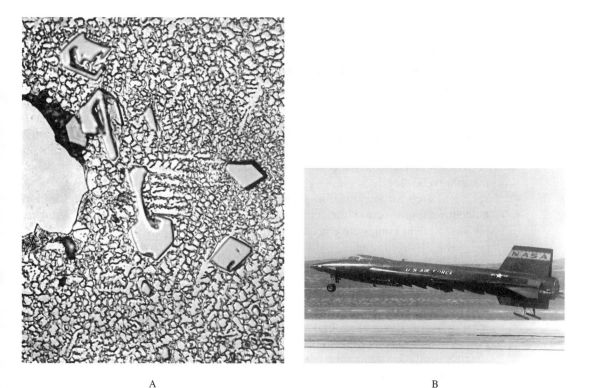

A   B

Figure 10.22   (A) A metallograph (by Todd Leanhardt). (B) The X-15 rocket-powered research plane. (Courtesy of NASA)

The two major categories of metal alloys are **ferrous alloys** and **nonferrous alloys.** Ferrous alloys use iron as the main constituent. The great use of ferrous alloys is due to the availability of iron ore, relatively inexpensive mining and fabrication processes, and the versatility of the materials produced. Some typical application areas are automobiles, pipes, and all sorts of large structures, such as buildings and bridges. Disadvantages of ferrous alloys include low strength-to-weight ratios compared to certain nonferrous alloys, high corrosion, and, sometimes, the fact that they are good electrical and heat conductors.

Superalloys, metals demonstrating extremely good performance at high temperature, have base metals of nickel, cobalt, aluminum, copper, magnesium, titanium, or iron (e.g. stainless steels). All except the stainless steels are **nonferrous superalloys.** TAZ-8A (Figure 10.22A) is a nickel-based superalloy with a relatively high percentage of aluminum to improve its oxidation resistance and high-temperature strength. The superalloy can take many forms (like the X-15 nose cone in Figure 10.22B), while offering high-temperature strength, oxidation and abrasion resistance, and exceptional thermal-fatigue resistance. Tests have shown that coating steel rolls with TAZ-8A increases resistance to thermal fatigue during temperature-cycling applications by 300 to 400%.

Platinum, also used for superalloys, is used to manufacture such items as **thermocouples,** capable of withstanding the temperature of molten glass, and spark plug electrodes with superior electrical conductivity. Platinum is considered a precious metal due to its chemical inertness, high melting temperature, and excellent electrical conductivity.

Manufacturability must also be considered. For example, **powder metallurgy** (P/M) is a manufacturing technique involving the compaction of powdered metal. P/M is used to produce alloys of materials having such high melting points that they are difficult to melt and cast. P/M materials are virtually nonporous and can maintain very close dimensional tolerances. Gears for lawnmower transmissions are made using P/M technology.

Research for improved high-performance metal alloys will be of great importance through the turn of the century. Companies involved in this research should enjoy an excellent competitive position.

The PMP Program, originated and supported by GE Superabrasives (Worthington, Ohio), is an international alliance of technical and community colleges and universities. Educators from these institutions learn the theory and application of superabrasive machining and grinding so that they can incorporate this technology into their curriculums. Member schools make superabrasive technology available to local industry. The goal of the PMP Program is to improve American manufacturing productivity and competitiveness. Training vocational and technical educators will ultimately produce a generation of students with knowledge of advanced superabrasive technologies. See Appendix A for contact information.

Ceramics

What comes to mind when you hear the word *ceramics*—bricks, fine china vases, or plumbing fixtures? These products are hard and brittle, but their desirable properties include lightness, attractiveness, and resistance to marring. They also do not rust, are resistant to most chemicals, and will stay the same over indefinite periods of time.

Ceramic materials are made by shaping and then firing a nonmetallic mineral, such as clay, at a high temperature. The different types of materials used for ceramics are abundant and readily available. Ceramics such as silicon nitride are known as functional ceramics as contrasted with clay, which is a structural ceramic. Functional ceramics are formed into parts of any shape such as cutting tools, valves, and bearings. The process used to produce these parts is known as superplastic forming. **Superplastic forming** presses material at high temperatures into almost any shape. Superplastic forming of metals has been around for a long time (see powder metallurgy, mentioned previously) but the process for ceramics is just now being developed. In some cases, ceramic materials are combined with metals (Figure 10.23).

New ceramic materials are already being used in the automotive industry. These materials have improved performance characteristics, including improved strength

Figure 10.23 Materials scientists use this apparatus to develop a new material called cermet—a combination of ceramics and metal. Cermet could be used in engine parts, cutting tools, and disk drives where wear resistance is desired. (Courtesy Lawrence Livermore National Laboratory)

and hardness (especially at high temperatures), and much better resistance to corrosion. They are much cheaper to manufacture than metals. The main problem in manufacturing is to keep out foreign materials such as water. Someday automobile engines may be entirely ceramic, making these engines cheaper, lighter, and more efficient than today's metal engines.

However, the most important future application for ceramics is **high temperature superconductivity (HTSC).** At fairly low temperatures, electrons in ceramic materials are coupled together in a way that allows them to flow more smoothly than they do in other materials (e.g., aluminum and copper). Electrons at higher temperatures (above $-300°F$ [88.6 K] in most materials) move randomly, wasting energy as they collide with atoms and impurities. High temperature superconductors, made of ceramics, are effective above a temperature of $-100°F$ (200 K). The latest research has

Figure 10.24 New transportation systems using high-temperature superconductivity will enjoy zero-frictional forces between levitated transported objects and a track.

moved us closer to the long-sought dream of room-temperature superconductivity. French scientists have reported superconductivity at $-10°F$ (250 K).

Ceramic superconductor materials are used in the magnetic levitation depicted in Figure 10.24. By now we have all heard of the breakthroughs this phenomenon could yield; for example, in the field of transportation where elevated trains move at fantastic speeds with zero friction.

Another application is to float manufactured products, usually in cartons, above a flat magnetic path. Levitated by ceramic superconductors inside the container, a linear motion built into the track effortlessly moves the cartons down the frictionless track. Magnetic levitation does not depend on gravity, so the conveyor system could be mounted in a ceiling to save space, or be effectively used in manufacturing systems in outer space. With no external moving parts the conveyor would not generate dirt, so it may be useful in clean rooms where semiconductor materials are made.

Interestingly, most superconductor research is in the area of electric power technology—generation, distribution, and end-user applications—such as the fabrication of motor windings. Superconductor material must be made into fine wires, which in turn can be made into large current-carrying coils and conductors. The all-electric automobile could be one outcome of such research.

Semiconductors

The last section's discussion of photovoltaics (PVs) revealed the need for improved **semiconductor materials.** Much basic materials research remains, but after the research questions are answered, the challenge of manufacturing will still exist. No benefits accrue to society until the new materials can be produced cheaply and reliably. It is the inability to rid the materials and processes of impurities that plague this important industry.

One potential research strategy is to use silicon with higher impurity levels. Pure silicon costs up to $75 a kilogram, while metallurgical-grade silicon costs only $2. Texas Instruments and Southern California Edison are working on a joint venture that could more than halve the present cost of solar-generated electricity. Six years of research has resulted in a manufacturing method that uses inexpensive silicon in solar cells. The result may be installed cells that could produce electricity for less than $1 per kilowatt hour.

Composites

A **composite material** has two or more chemically distinct reinforcing materials. Composite materials are leading the way in new materials usage, and in many consumer items they are difficult to identify. The most common kind of roofing shingle, for example, is a composite of glass and asphalt.

Composites can be as strong as metals, yet lighter, cheaper, and rust-free. Composites may be found in such diverse items as furniture, automobiles (even engine parts), concrete, plastic piping, grinding wheels, tennis rackets, and space station solar array structures.

Reinforced plastic technology represents the fastest-growing area in the field of composite materials. *Fiberglass-reinforced plastic* (FRP) is a composite, for the glass fibers are incorporated into a resin matrix. The resin is an organic product that can occur naturally or can be synthetic, and the glass fibers are used to reinforce the resin. Over 90 percent of reinforced plastics are FRP materials.

Composite materials are designated as fibrous, particulate, or laminar. These areas can be further described as follows:

FIBROUS: Fibers are embedded in a continuous matrix. The fibers may be natural or synthetic, organic or inorganic (Table 10.5).

PARTICULATE: Particles are embedded in a continuous matrix. Portland cement concrete is the most common particulate composite. The properties of the concrete are enhanced by the addition of metal or glass reinforcement and by modification of the matrix with polymeric (forming molecules from simpler molecules of the same kind) binders.

LAMINAR: Sheets are bonded together in a sandwich construction. The area between two continuous sheets is often honeycombed when strength and flexibility are required.

Manufacturing composites is difficult, strongly linking the technology to advancements in computer technology. Blending the materials in the correct ratios, while simultaneously controlling other factors such as temperature and viscosity, is best accomplished by computer control. Also, a computer can model the strength of a new material to determine load bearing characteristics when cut or shaped into various configurations.

Table 10.5   Examples of Fibers Used in Fibrous Composite Materials

	Natural	*Synthetic*
Organic	Cellulose Jute Pulp	Aramid Polyester Rayon
Inorganic	Asbestos*	Steel, aluminum Carbon graphite Glass

*This material has been known to cause lung irritation and is possibly carcinogenic.

Major areas where composites are used include

AIRCRAFT AND AEROSPACE: The high strength-to-weight ratio of composites make them the material of choice for many applications in the aerospace industry. The space flight of *Voyager* was made possible by a composite material (plastic reinforced with graphite). Entire bodies on future aircraft may be filament-wound reinforced plastic.

APPLIANCES: Electronic cabinets, refrigerators, washer-dryers, and small motor casings are fabricated from plastic composites. This is due to the ease with which these materials can be molded into complex shapes in one manufacturing operation.

ARCHITECTURAL: Residential and commercial construction will use composite materials increasingly in modular units as composites gain acceptance. Composites are already replacing traditional materials for flooring, plumbing, wall coverings, and lighting fixtures.

FURNITURE: Reinforced foams enable designers to use composites in living room and bedroom furniture. School and recreational seating consists of plastic composites.

TRANSPORTATION: Efficient transportation also results from high strength-to-weight ratios. Composite materials are being increasingly used in autos (Figure 10.25), trucks, and boats. Bridges are being built and roads quickly resurfaced by environmentally friendly composites developed by NASA.

Parts consolidation is a major advantage of composites. Complex shapes can be molded into one piece, eliminating the necessity of welding many joints.

Areas of interest and important skills in working with composites include

- identifying different types of materials and their bonding agents;
- knowing the compatibility of one class of materials with another (this is an important area the technician will be counted on to know—e.g., whether or not to use an off-the-shelf lubricant with a certain type of polymer);
- building and operating complex machinery such as extruders and chemical blending systems; and
- testing materials for strength, ductility, and resiliency.

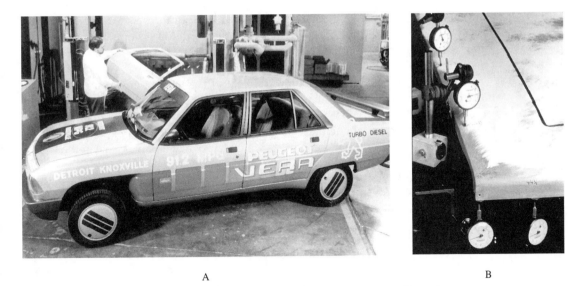

A   B

Figure 10.25 (A) Composite materials are used in automobile bodies, (B) requiring new manufacturing processes. (Courtesy Owens-Corning Fiberglas)

Technicians and technologists will play crucial roles in the manufacture of composites as well as other new materials during the end of this century and the beginning of the next. Materials technicians will be needed in the following technologies: chemical engineering technology (ChET), mechanical engineering technology (MET), and computer engineering technology (CpET).

10.5 Our Environment

We cannot close our discussion of technology without considering its potential for improving or destroying the environment we share with all other life forms. Spaceship Earth is a finite planet with limits that may easily be reached by our rapidly growing human population. In some cases, limits have already been reached.

Most of us can name many of the major ills of our environment: suspected global warming, due to increased air pollution; deforestation, due to demands for more crop space and acid rain; loss of topsoil, due to poor land-management practices such as overworking present crop space; scarcity of safe drinking water, due to ground water pollution and increasing demands; and suspected depletion of the ozone layer, due to the use of chlorofluorocarbons.

The primary reason for most of the ills listed above, is the human population explosion. In the five years since the Rio de Janeiro Earth Summit in 1992, the earth's population has grown by another half a billion. The developing world is building new manufacturing and power plants, which will add yet more carbon-dioxide (CO_2) to our atmosphere. Expanding economies will squeeze out forests that clean our air of harmful gases. More people require extensive transportation systems, large industries, and many products, resulting in contamination of precious air and water and an increased demand on other limited natural resources.

Manufacturing processes require materials that become environmental waste. The cutting oil used in machine tool operations is an example. Most cutting oil is now recycled, but not long ago it was dumped in collecting ponds that contaminated the ground water. Industrial chemicals in ground water have led to long-term or permanent evacuations of entire communities, which happened at Love Canal in the United States and Seveso in Italy.

Technology has obviously contributed to the ills of our environment. Perhaps less well known are the ways technology can improve our environment. The challenge for the technologist and the focus of this section is to learn

- how to improve the environment through technology, and
- how to protect the environment from the misuse of technology.

Improving our Environment

You should be aware of the ways technology has already improved our environment. The computer has saved tons of paper products by keeping data stored in digital formats. Drawings, text, and numerical information can all be recorded electronically using 1/1000th of the space required for paper. To further illustrate, a single CD-ROM contains information that would normally be written on 700 kg (1540 lb) of paper. Data needed in the field can be accessed much more easily with a computer than with the reams of paper required previously (Figure 10.26).

Technology *harmed* our environment with the development of automobiles and other vehicles that required leaded gasoline for more efficient engines. Technology has now improved our environment through the development of unleaded gasoline. This single act has reduced lead levels in the atmosphere more than seven times between 1967 and 1989, according to ice samples collected in Greenland. The U.S. Environmental Protection Agency (EPA) has recorded a dramatic decrease of 88% in atmospheric lead levels in the United States.

Technologists have improved the environment by eliminating waste in manufacturing processes. Increased automation makes processes more efficient. Improved quality control will reduce scrap and rework.

Improved instrumentation is not only automating manufacturing processes, but also measuring the changes in our environment. Sensors placed in strategic locations, calibrated and maintained, measure what drifts up the smokestack, gets trucked out of the company gates, or flows into the sewer. More funds must be made available to adequately monitor contaminants that enter our air and water. We will need to prevent pollution, not just control it. **Pollution prevention** techniques, also requiring advanced instrumentation, will change raw materials or production techniques to be more compatible with natural materials and processes. The chemical revolution has resulted in new forms of waste that may be hazardous.

Figure 10.26 The environmental technician accesses data in the field. (Courtesy International Business Machines Corporation. Unauthorized use not permitted.)

New research will result in better methods for identifying cancer-causing contaminants, so they may be controlled or eliminated (Figure 10.27).

Technologists may work through their professional associations, such as the Instrument Society of America, to encourage more government and private, non-profit support for environmental instrumentation. The trend for the use of environmental instrumentation is up, with over $2 billion in pollution analyzer sales in 1995.

We need to make better use of the oceans and outer space. The oceans of the earth make up more than 70 percent of its surface. We cannot neglect them. Food products may be harvested from the oceans, but overfishing of popular large fish could eliminate many species. Smaller animals and plants of the sea may provide better food sources in the future. Seawater contains every mineral found on land; for example, it holds tons of gold. The challenge is to learn how to more efficiently extract the minerals we need without damaging the environment. Mining of the ocean's floor will yield valuable minerals, such as manganese, copper, nickel, and cobalt.

The moon also offers extensive mineral resources. Chemicals such as hydrogen, carbon, and nitrogen can be extracted from the lunar soil by using the sun's energy to

Figure 10.27 This mass spectrometer can study the effects of potential cancer-causing chemicals on human cells. (Courtesy Lawrence Livermore National Laboratory)

heat the surface to about 700°C. We must not let the drive to explore other planets deter us from developing our closest extraterrestrial resource, the moon.

One way to find out how people could live in space or underwater is to create an artificial environment such as Biosphere II, which is located in Oracle, Arizona, near Tucson. Biosphere II (Biosphere I is the earth) helps us to understand our entire earth's environment. Unique features of this totally independent environment include:

- a double-laminated glass-and-steel geodesic frame;
- the recycling of air, water, and waste in the biosphere;
- a rain forest with an 85-foot-high waterfall and a 25-foot-deep ocean; and
- dome shaped "lungs," totaling nearly an acre, connected by a tunnel to allow a buildup of gases during the daytime so the glass panels will not blow out.

The creators of Biosphere II have sought more than 50 design patents for new technologies.

Protecting Our Environment

Forward-looking industries are working to become *greener.* This means they are putting their employees, along with environmental consultants, to work on every industrial process in order to find out how to make the process more efficient with nature. Solar energy (Section 10.3) may be used someday to reduce pollution and use less oil. Materials that end up in waste bins may be used to make other products. New products can be designed to be more ecologically friendly. For instance, carpet for business use can be cut into small squares instead of being made to cover large areas. The small squares can be independently replaced when worn and more easily recycled into new carpet.

The *green industries* will be more attractive to customers and investors. The technologist will be a major player in these industries. The technologist is often the person most familiar with the day-to-day operation of a business or factory. The technologist will often be the person recognizing practices that adversely affect our environment, such as when an industry is dumping harmful chemicals into a river or lake. Technologists must be aware of what practices will damage the environment so they can inform management of questionable practices. The difficult problem is to decide what to do when such reports are *not* followed by corrective action.

Article I of the *ISA Code of Ethics* (Section 3.4), reads: "Members shall hold paramount the safety, health and welfare of the public in the performance of their duties, and shall notify their employer or client and such other authority as may be appropriate where such obligations are abused." This ethic, to accept responsibility for the health and safety of the public, is often one of the most difficult to honor. How many of us would be willing to put others before ourselves, to risk our jobs by reporting a manufacturing process or business practice that is harming our environment? Yet we all agree that sometimes such drastic action is necessary. Only a few will have the courage to be the *whistle-blowers* of our society. They will be heroes to future generations.

As technologists we have a responsibility to ourselves, to others, and to future generations. We cannot neglect environmental concerns. Technology can injure or heal our precious environment. We must work to decrease technology's ability to injure our environment and increase its ability to heal it.

10.6 Meeting the High-Technology Needs of the World— A Challenge

We are entering an exciting era. The body of scientific knowledge is estimated to double about every decade. This wealth of knowledge, resulting in countless new products and equipment, offers humankind improved lifestyles and a quality of life unknown in this century. However, sophisticated equipment could be useless, or even dangerous, without people who understand and can implement the new technology.

Technologists hold the key to how quickly these innovations will be implemented. The knowledge base advances so fast that most technology becomes out-of-date in five to seven years. In electronics, it's two to three years. To remain competitive, governments must provide quality education for new technicians. Business and industry must continue to invest in training throughout technicians' careers. It is the technical work force that must install and maintain safe and effective high-technology systems.

Manufacturing output as a percentage of the gross national product in the U.S. has remained about the same over the last decade. Manufacturing employment is being reduced by automation; however, employment potential for technicians and technologists in manufacturing will remain favorable. The hard-technology industries continue to employ the same numbers of engineers and technicians despite declining employment of skilled and semiskilled production workers. High-technology industries will continue to hire well-prepared technologists.

Computer-aided manufacturing has made even assembly-line jobs more complex. The need for technicians will not be met for years by technical school graduates. Many industries are working directly with two-year technical colleges to begin **cooperative education** programs eventually leading to a degree in one of the manufacturing-related technologies. Whether you are employed in older, but vastly modernized basic industries, or in the newer information industries, you must possess a balance of technical knowledge and practical experience. Cooperative education programs will provide this balance.

Technicians and technologists will be needed in the fast-growing **service industries.** In some service-related areas, technicians and technologists are in short supply. The Motor Vehicle and Manufacturers Association, for instance, feels recruiting and retraining "a new breed of technician who can utilize new diagnostic equipment and accurately interpret data to pinpoint the problem with an automotive system" is one of the industry's major concerns. The association cites the need to, "find and train a minimum of 32,000 technicians a year through at least 1995, an effort that has fallen far short of the mark to date."

Service industries include government, nonprofit, and international organizations. Technologists in these service positions will implement emerging technologies and transfer those technologies to developing nations.

Technology transfer is a pressing need in the 1990s and beyond. Countries successfully implementing new technologies will be more in control of their economies and their destinies. Countries not in control of new technologies will have weak economic growth and will not be able to plan and implement their futures. Emerging democracy will be threatened in some of these countries and chaos may continue in many developing nations. These developing nations will need to import expertise. Some of you, entering a career in the technologies, will undoubtedly be of service to these nations. From a humanistic standpoint, we cannot deny that we have the capacity to reduce human suffering by use of technology, suffering that we have almost forgotten because it is not so evident in the industrialized nations. Many of you may travel to distant places to teach others how to use modern technological systems that will improve the human condition in those areas.

Being active in your professional associations will keep you aware of current challenges for technology and how to best meet the overall needs of society. Strive to be well rounded. Cultivate different interests, support the most able political candidates, and remember that people all over the world desperately need and depend on the implementors of the technological community, and you.

Problems

1. Research, list, and describe in one paragraph or less three areas of specialization that may be of interest to you. See a partial listing of specializations in the introduction to this chapter.

Section 10.1

2. What is meant by "the exponential growth in computer technology" (see Figure 10.1)?

3. Define *cybernetics* and *heuristics* (use a good dictionary). How do these concepts relate to artificial intelligence (AI)?

4. Describe an application for an expert system. Explain, in a general fashion, what diagnostic steps the program might take.

5. Survey the office support staff at your college or place of employment, in order to ascertain how computers are used in the office environment. Prepare a brief report summarizing your findings.

6. Discuss in two or three paragraphs the difference between an analog and digital signal. Also comment on the quality of each in telecommunications systems.

7. Research and briefly describe ADSL and ISDN.

Section 10.2

8. List the major components of a robot.

9. Describe the function of each of the major robotic components.

10. Describe the different types of robot controllers.

11. Design and sketch a robot with four degrees of freedom.

Section 10.3

12. List the three major areas of current advances in optical research.

13. If a photovoltaic's efficiency is defined as $/watt, by what percentage have efficiencies increased since the early 1970s?

14. See Table 10.4, part 3. With nonsilica-based, fiber-optic cable, how many amplifiers may be needed in a trans-Pacific telephone cable to Japan? (The distance to Japan is 6 000 miles.)

15. One of the fastest growing technical areas in the future will be optical systems. Research one magazine article on photovoltaics, lasers, or fiber-optic systems. Copy the article for submission to your instructor, or write a brief synopsis of the article (instructor's determination).

Section 10.4

16. List the main categories of advanced materials discussed in this section.

*17. Describe one type of advanced material used in your home.

*18. Discuss the advantages of the use of titanium in golf clubs.

---

*Challenging problem.

19. List some uses for ceramic materials, especially the new, high-technology ceramics.

20. Report on the benefits of superconductivity research.

Section 10.5

21. List the two challenges for the technologist that are the focus of this section.

*22. Research and describe one environmental problem. Discuss what must be done to eliminate or at least ease the problem.

*23. Describe what the author means by a *green industry.*

24. Research and copy one journal article addressing a high-technology solution to an environmental problem.

Section 10.6

25. Research and describe the needs of one developing country. Explain how technologists might someday help meet those needs.

26. Write a position paper—take the position or argue for—the need for technologists to be well rounded. Explain how you can keep up with rapid change in your technology, yet take time to keep current with local and world affairs and enjoy your hobbies.

27. After touring a local industry, write a tour report. The report should be one to two pages in length. Include a brief, informal overview of the company visited, and address the following topics:
    a. the name and trademark of the company visited.
    b. the mission of the company.
    c. a description of one of the processes or a part of the product.
    d. a major quality control (QC) test conducted.
    e. a description of the product's packaging.

The conclusion should summarize your impressions of the company's ability to fulfill its mission.

28. Using the *Occupational Outlook Handbook* or the *Dictionary of Occupational Titles,* list five career positions for one of the following selected technologies: robotics, optical systems, materials, or the environment.

Selected Readings

Engelberger, Joseph F. *Robotics in Service.* Boston: MIT Press, 1989.

Fuller, James L. *Robotics: Introduction, Programming, and Projects.* Upper Saddle River, New Jersey: Prentice Hall, 1991.

Hawken, Paul. *The Ecology of Commerce.* New York: Harper Collins, 1993.

Materials Science and Engineering for the 1990s. Washington, D.C.: National Academy Press.

Naisbitt, John, and Patricia Auberdene. *Megatrends 2000.* New York: Avon Books, 1990.

Pimental, Ken, and Kevin Tiereira. *Virtual Reality.* New York: McGraw-Hill, 1993.

Technology 2001: The Future of Computing and Communications. Boston: MIT Press, 1991.

*Challenging problem.

Appendix A

Professional Organizations in Engineering*

American Society of Heating, Refrigerating, and Air-Conditioning Engineers (ASHRAE). 1791 Tullie Circle, NE Atlanta, GA 30329. 800.5-ASHRAE

American Society for Quality (ASQ). 611 E. Wisconsin Ave., Milwaukee, WI 53201-3005. 800.248.1946

Educational Society for Resource Management (APICS). 500 W. Annandale Rd., Falls Church, VA 22046. 800.444.2742

Institute of Electrical and Electronic Engineers (IEEE). 345 E. 47th St., New York, NY 10017. 212.705.7900

Institute of Industrial Engineers (IIE). 25 Technology Park, Norcross, GA 30092. 770.449.0460

International Society for Measurement and Control (ISA). 67 Alexander Dr., P.O. Box 12277, Research Triangle Park, NC 27709. 919.549.8411

National Association of Business and Educational Radio (NABER). 1501 Duke St., Alexandria, VA 22314. 800.759.0300

National Fluid Power Association (NFPA). 3333 N. Mayfair Rd., Milwaukee, WI 53222-3219. 414.778.3361

National Institute for Certification in Engineering Technologies (NICET). 1420 King St., Alexandria, VA 22314-2794. 800.787.0034

Partnership for Manufacturing Productivity (PMP). G.E. Superabrasives, 6325 Huntley Rd., Worthington, OH 43085. 704.684.1988

Society of Manufacturing Engineers (SME). P.O. Box 930, One SME Dr., Dearborn, MI 48121-0930. 800.733.4763

Society of Plastics Engineers (SPE). 14 Fairfield Dr., P.O. Box 0403, Brookfield, CT 06804-0403. 203.775.0471

Society of Women Engineers (SWE). 120 Wall St., 11th Floor, New York, NY 10005. 212.509.9577

*A selected listing of those most responsive to technicians. Many of them support student sections.

Appendix B

Algebraic Rules

The following summary of "selected" rules for algebra is offered to enable you to avoid mistakes when performing operations with signed numbers and solving for literal factors. *You must be able to list these rules from memory in order to be competent in algebra* and algebraically related operations.

In Chapter 4 signed numbers were introduced. Listed below are the rules for signed numbers when adding, subtracting, multiplying, and dividing.

Algebraic Addition

Like signs:

> To add two numbers with like signs, add their absolute values and attach the common sign.

Unlike signs:

> To add two numbers with unlike signs, subtract the smaller number from the larger. The result carries the sign of the larger.
>
> *Examples:* $(-2) + (-8) = -10$ and $(-4) + (+7) = +3$

Algebraic Subtraction

Change the sign of one of the numbers and algebraically add. Because algebraic subtraction represents a distance between two numbers on the number line, the absolute (unsigned) value is often used for the result.

> *Examples:* $(-5) - (+2) = -5 - 2 = -7$ or $|7|$
>
> and
>
> $(18) - (-8) = 18 + 8 = 26$
>
> (i.e., there are 26 [absolute value] units between 18 and -8 on the number line)

Algebraic Multiplication and Division

If the signs of the two numbers are alike the result is positive. If the signs are opposite the result is negative.

Examples: $(-2) \times (-3) = 6$ and $(2) \div (4) = 0.5$

$(6) \times (-2) = -12$ and $(-12) \div (3) = -4$

Rules of Exponents

Rule	*Example*
$a^0 = 1$	Any base to the zero power is one
$a^x \times a^y = a^{x+y}$	$3^2 \times 3^4 = 3^6$
$a^x \div a^y = a^{x-y}$	$3^2 \div 3^4 = 3^{-2}$
$a^{-x} = \dfrac{1}{a^x}$	$3^{-2} = \dfrac{1}{3^2} = 0.111$
$(a^x)^y = a^{xy}$	$(3^2)^4 = 3^8$
$\left(\dfrac{a^x}{b^z}\right)^y = \dfrac{a^{xy}}{b^{zy}}$	$\left(\dfrac{2^2}{3^3}\right)^4 = \dfrac{2^8}{3^{12}}$
$\sqrt[y]{a^x} = a^{x/y}$	$\sqrt[3]{3^2} = 3^{2/3}$

Cross Multiplication

$$\frac{a}{b} \diagdown\!\!\!\!\diagup \frac{x}{y} \ \text{or}\ ay = bx \ \text{or}\ y = \frac{bx}{a} \ \text{or}\ x = \frac{ay}{b}$$

Cross multiplication works only for monomials (no + and − signs). Factors may be moved across the equal sign by moving them (1) from numerator to denominator or (2) from denominator to numerator.

The cross multiplication principle shows how algebra (and the use of literal factors) helps us to *think in patterns.*

The Reverse Order Solution

Solving equations for a specific literal factor in an equation (occurs only once) may be facilitated by using a reverse order solution set (or by unwinding the equation). For example:

$$\sqrt[3]{X^2 + 2} - 7 = -4$$

If you knew the X was equal to 5 ($X = 5$), you would prove the equality by

1. squaring the X value ($5^2 = 25$),
2. adding the 25 to 2 ($25 + 2 = 27$),
3. taking the cube root of 27 ($\sqrt[3]{27} = 3$), and
4. adding the 3 and the -7 to obtain the -4; thus proving the equality ($-4 = -4$).

To solve for X as an unknown, merely *follow the above steps in reverse order.*

1. Take the 7 to the other side by changing its sign and adding it to the -4 (answer = $+3$).
2. Cube both sides to obtain $X^2 + 2 = 27$ ($3^3 = 27$).
3. Move the 2 to the other side (change its sign) and—after algebraically adding— the result is $X^2 = 25$ ($27 - 2 = 25$).
4. Take the square root of both sides to obtain $X = 5$ ($\sqrt{25} = 5$).

Illegal Operations

It is just as important to know what you cannot do. Below are mistakes beginning algebra students tend to make with polynomials ($\neq$ means "does not equal").

Illegal Operation	Reason
$\dfrac{X^2 + 4}{3X + 2} \neq \dfrac{X^2 + 2}{3X + 1}$	Only monomial factors may be cancelled.
$(X + 2)^3 \neq X^3 + 2^3$	The $(X + 2)$ must be multiplied by itself three times

to obtain

$$X^3 + 6X^2 + 12X + 8$$

If $\dfrac{1}{R_T} = \dfrac{1}{R_1} + \dfrac{1}{R_2}$

Both sides must be inverted, and $R_1 R_2$ is the common denominator:

then $R_T \neq R_1 + R_2$

$$R_T = \frac{1}{\dfrac{1}{R_1} \dfrac{1}{R_2}} = \frac{1}{\dfrac{R_2 + R_1}{R_1 R_2}}$$

and, after inverting

$$R_T = \frac{R_1 R_2}{R_1 + R_2}$$

Appendix C

Trigonometry

This appendix will show you how to solve for any unknown in a right triangle, given any two other parameters. (Remember, the 90-degree angle is always known.)
You must also know

1. the Pythagorean theorem ($c^2 = a^2 + b^2$), and
2. the complementary rule (in any right triangle the two angles [other than the right angle] always add up to ninety degrees, or $A + B = 90°$).

Figure C.1 shows a right triangle in standard position with legs labeled opposite and adjacent for angles A (in A) and B (in B). The longest side, the side opposite the right angle, is called the hypotenuse.

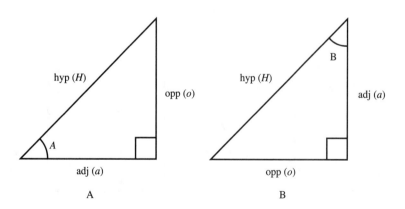

Figure C.1　The right triangle in standard position.

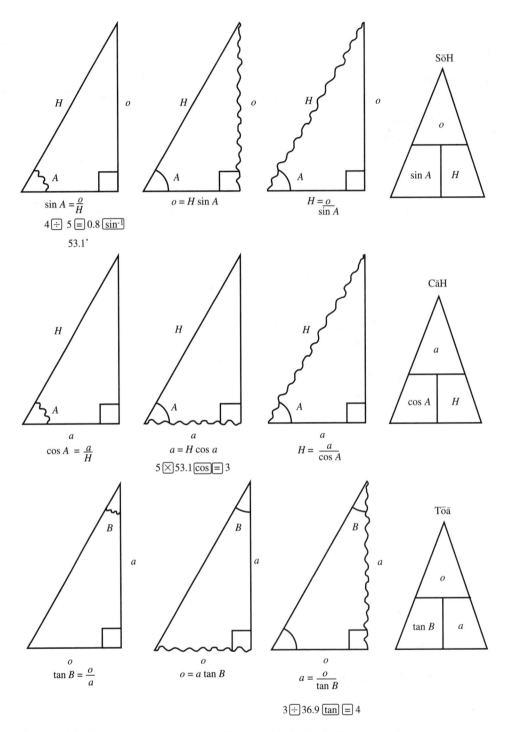

Figure C.2 In this chart, the unknown (item to be found) is illustrated by a wavy line. Find the illustration needed, select the correct equation, and solve for the unknown. The calculator solutions solve for a 3-4-5 triangle (base = 3 and vertical side = 4).

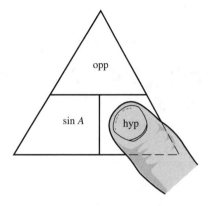

Figure C.3   A solutions template for the sin θ relationship.

This appendix should be used only until you master the concepts of algebra. The equilateral triangles at the side of Figure C.2 will no longer be necessary once you have acquired the powerful tools of algebra. But for now, to solve for an unknown, simply cover the unknown on the equilateral triangle with your finger (Figure C.3). The other two literal factors are seen in the correct mathematical relationship. In this case,

$$H = \frac{\text{opp}}{\sin A}$$

The ancient village of SOH-CAH-TOA will help you remember the three trigonometric relationships:

$$\sin A = \frac{\textbf{opp}}{\textbf{hyp}}; \ \cos A = \frac{\textbf{adj}}{\textbf{hyp}}; \ \tan A = \frac{\textbf{opp}}{\textbf{adj}}$$

Appendix D

Beginning BASIC Commands and Reserved Words

Command/Statement	Action or Example
ASIN, ACOS, ATN	Arc functions (trigonometry)
BEEP	Warns the program user of an invalid entry
CLS	Clears screen
END	Ends the program's execution
GOSUB (line #)	Control advances to the argument's line number, then returns to the line number immediately following the GOSUB command, with the RETURN command
GOTO (line #)	Control advances to the argument's (parentheses) line number—e.g., for AND command: IF N > 0 AND Y < 50 GOTO 350 (if both inequalities are satisfied the program advances to line 350)
IF- ...... THEN	*Example:* IF X = 5 THEN GOSUB 120
INPUT	Causes program to stop and accept input from operator; a question mark (?) follows the INPUT command
LIST	Lists program lines on the screen
LLIST	Lists program lines on the printer
LOAD	*Example:* LOAD "POLTOREC" loads the program from floppy disk to the computer's RAM
NAME	Changes the name of a file; *example:* NAME "POLTOREC.BAS" AS "COORCON.BAS" (coordinate conversion in BASIC)
NEW	Erases everything in memory
NOT	*Example:* IF NOT N > 0 THEN RETURN
OR	*Example:* IF N > 0 OR Y < 50 GOTO 45
PRINT	Executes an output to the printer peripheral
REM	Remark; program notes not executed in the RUN mode
RETURN	See GOSUB command

RUN	Runs the current (currently loaded; see LOAD) program
SAVE	Saves a program with a designated name; *example:* SAVE "POLTOREC"
SYSTEM	One of the commands allowing you to exit the BASIC mode
TAB(N)	Line characters move (tab) to the screen position designated as N; N must be a value between 0 and 80 ($80 \geq N > 0$); also see use of LOCATE command in Chapter 9's program example (POLTOREC)

Appendix E

Selected Computer Languages*

ADA Programs written for military weapons and space systems are most frequently written in ADA. ADA is an easily structured language. Different programmers, experts in their fields, can write independent programs (modules), which can then be easily combined into a total system program. The computer-controlled space shuttle makes use of over one million lines of ADA programming code.

C Developed by Bell Laboratories, C is used mostly in writing system software. One advantage of C is that programmers can easily control computer hardware. C also lends itself to structured programming (see ADA above) and is increasingly used to write new application programs.

COBOL COBOL stands for *common business oriented language*. Used in coding business data-processing problems, it is an excellent file-handling language. The business world makes extensive use of COBOL, and—even though it is a wordy and somewhat confusing language—it will remain in use for some time.

FORTRAN FORTRAN (*formula translator*) is the oldest of the high-level programming languages, and it contains an impressive library of engineering subprograms. It is used, mainly in research and development, to solve mathematical problems. Once the most popular language in science and engineering, its use is now declining because of a lack of input/output commands. It is interesting to note that BASIC was originally designed to aid students to learn FORTRAN.

LISP LISP is an example of a symbolically oriented (rather than a numerically oriented) language. Symbolically oriented languages are useful in writing artificial intelligence (AI) programs and expert systems. C is also being considered in writing AI programs.

---

*See Chapter 9 for descriptions of machine, assembly, and BASIC languages.

Pascal   Pascal and Modula-2 (an improved version of the original Pascal) are used mainly to teach structured programming to students. These languages offer strong data structures, with succinct language requirements.

PROLOG   Based on formal logic (a unique mechanism for implementing the program), PROLOG is being considered—especially by the Japanese—as the native language of the fifth-generation computer.

Appendix F

Glossary of Abbreviations and Acronyms

A listing of the most frequently used technical abbreviations and acronyms. The technician and technologist must be familiar with most of the list; they are universally used terms in the applied technologies and in the associated literature. The alphabetical listing is by the abbreviations, and not by the spelling of the first word. For definitions, consult the index for page numbers.

A/D: Analog to Digital converter (pronounced A to D)
ADSL: Asymmetric Digital Subscriber Line
AI: Artificial Intelligence
ALS: Algebraic Logic System
CAD: Computer-Aided Design
CADD: Computer-Aided Design and Drafting (also CAE, Computer-Aided Engineering)
CAM: Computer-Aided Manufacturing
CD-ROM: Compact Disk with Read-Only Memory
CID: Career Information Delivery (CID) system
CIM: Computer-Integrated Manufacturing (pronounced sim)
CISC: Complex Instruction Set Computer (pronounced sisc)
CNC: Computer-Numerical Control
CPU: Central Processing Unit
CRT: Cathode Ray Tube
D/A: Digital to Analog converter (pronounced D to A)
DCS: Distributed Control System
DD: Double Density (floppy disk)
DOS: Disk Operating System
DVD: Digital Video (or Versatile) Disk
FCC: Federal Communications Commission
FMS: Flexible Manufacturing System
FRP: Fiberglass-Reinforced Plastic
HD: High Density (floppy disk) (also Hard Drive)
HTSC: High-Temperature Superconductivity
HVAC: Heating, Ventilating, and Air Conditioning
IC: Integrated Circuit

ISA: Industry Standard Architecture (pronounced IS-uh)
ISDN: Integrated Services Digital Network
JIT: Just In Time manufacturing
LAN: Local-Area Network
LASER: Light Amplification by Stimulated Emission of Radiation
NC: Numerical Control (also Network Computer)
NEC: National Electrical Code
PC: Personal Computer
PCI: Peripheral Component Interconnect
PCMCIA: Personal Computer Memory Card International Association
PLC: Programmable Logic Controller
PV: Photovoltaic cell
QC: Quality Control
RAM: Random Access Memory
R&D: Research and Development
RISC: Reduced Instruction Set Computer (pronounced risk)
ROM: Read-Only Memory
RPN: Reverse-Polish Notation
SCSI: Small Computer System Interface (pronounced SKUH-zee)
SI: International System of units
SPC: Statistical Process Control
USCS: U.S. Customary System

Answers to Odd-Numbered Problems

Chapter 1

1. Requires personal library research.
3. Requires individual response.
5. Requires individual response.
7. Communicator, implementor, cal and tester, manufacturing engineer.
9. Requires personal research.
11. Occupational satisfaction, availability of employment, salary potential.
13. An increase in automation requires complex machinery. Technicians or technologists with a quality education are needed to fabricate, maintain, or repair such equipment.
15. Requires personal library research.
17. Requires personal research.
19. Requires internet research.

Chapter 2

1. Chemical Engineering Technician, Civil Engineering Technician, Electronic Engineering Technician, Computer Engineering Technician, Industrial Engineering Technician, Mechanical Engineering Technician.
3. Requires personal report.
5. Requires personal report.
7. Requires personal report.
9. Requires personal research.
11. Requires research in college catalog.
13. MET
15. ChET
17. CpET

19. MET (or CET)

21. EET

23. MET

25. CET

27. MET

29. CET

Chapter 3

1. Associate of Science in Engineering Technology—technician
   Bachelor of Science in Engineering Technology—technologist
   The central purpose of both programs is to get a job.

3. See Section 3.1, *The Curriculum and Succeeding in College Life*

5. Requires individual response.

7. Requires visiting college placement office and report.

9. Requires creation of new file. A large manila folder will be needed.

11. Requires plan of study.

13. Requires library research.

15. Requires library research.

17. 2.00

19. Recognize problem, define problem, brainstorm alternatives, determine consequences of alternatives, select best solution, implement solution, and evaluate with feedback loop to the beginning.

21. Requires individual problem solving.

23. Requires individual problem solving.

25. Requires individual response.

27. See Section 3.4's subhead, *Standardization,* and the American National Standards Institute (ANSI)

29. It is important to complete this assignment.

Chapter 4

1. Consult your instructor for recommended calculator dealers.

3.  A typical calculator's display could have these screen prompts

5. 1 $\boxminus$ 2 $\boxminus$ 1 $\boxplus$ 7 $\boxminus$ 5

7. 7 $\boxminus$ 6 $\boxplus$ 4 $\boxminus$ 5

9. 91 $\boxed{\times}$ 22 $\boxed{\times}$ 35 $\boxed{\div}$ 12 $\boxed{=}$ 5840

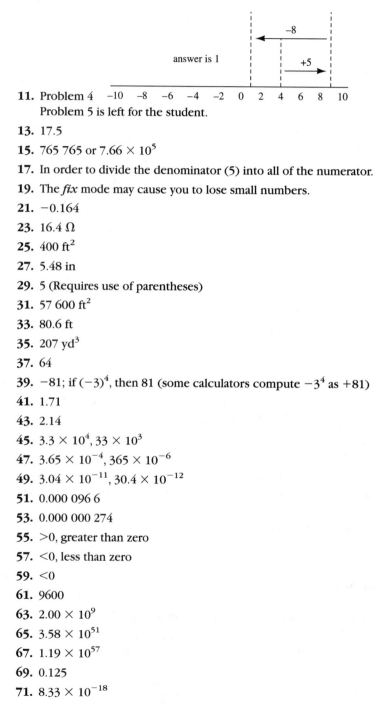

11. Problem 4
Problem 5 is left for the student.

13. 17.5

15. 765 765 or 7.66×10^5

17. In order to divide the denominator (5) into all of the numerator.

19. The *fix* mode may cause you to lose small numbers.

21. −0.164

23. 16.4 Ω

25. 400 ft^2

27. 5.48 in

29. 5 (Requires use of parentheses)

31. 57 600 ft^2

33. 80.6 ft

35. 207 yd^3

37. 64

39. −81; if $(-3)^4$, then 81 (some calculators compute -3^4 as +81)

41. 1.71

43. 2.14

45. 3.3×10^4, 33×10^3

47. 3.65×10^{-4}, 365×10^{-6}

49. 3.04×10^{-11}, 30.4×10^{-12}

51. 0.000 096 6

53. 0.000 000 274

55. >0, greater than zero

57. <0, less than zero

59. <0

61. 9600

63. 2.00×10^9

65. 3.58×10^{51}

67. 1.19×10^{57}

69. 0.125

71. 8.33×10^{-18}

73. 128

75. 279

77. $I = 40, 100,$ & $140 \, \mu A$

79. 1.05×10^6 bits ≈ 1 million bits (8192 bits = 1 KB of RAM) see Front Inside Cover.

81. $X \approx 1.75$ in $V \approx 91.9 \, \text{in}^3$

Chapter 5

1. **a.** B, f   **b.** SI, f   **c.** E, d   **d.** B, f   **e.** E, d   **f.** E, f
   g. E, d   **h.** SI, d   **i.** B, f   **j.** SI, f   **k.** SI, d   **l.** E, d
   m. E, d   **n.** SI, f   **o.** SI, f   **p.** SI, d   **q.** B, f

3. Refer to Section 5.1.

5. **a.** 8.89°C   **b.** −239°C
   c. −40°F (the only temperature where both scales read the same)
   d. 187 K   **e.** 1297 K

7. $I = \dfrac{E}{R} = \dfrac{6 \, V}{24 \, \Omega} = 0.25 \, A$

9. **a.** $\dfrac{1 \, \text{min}}{60s}$ **b.** $\dfrac{1 \, \text{sl}}{32.2 \, \text{lb}}$ **c.** $\dfrac{12 \, \text{in.}}{1 \, \text{ft}}$ **d.** $\dfrac{100 \, \text{cm}}{1 \, \text{m}}$ **e.** $\dfrac{1000 \, \text{mA}}{1 \, A}$

11. The 8.5 in., when converted to feet, yielded a rounded-off value of 0.708 ft. The actual conversion to 10 significant figures is 0.708 333 333 3. Rounding-off error makes soft conversion between different units too inexact to be practicable (e.g., converting a metrically designed automobile to U.S. customary units before manufacture).

13. 6160 ft/min.

15. 8.33 min.

17. 1. c   2. f   3. j   4. g   5. h   6. a   7. e   8. b

19. 50.3 cm, 503 mm

21. 101.3 kPa

23. micrometer

25. $\dfrac{0.0394 \, \text{in.}}{1 \, \text{mm}}$

27. 39.3 ft/min, 259 ft

29. 13 mm

31. bilateral tolerance in mm is $+0.0254, -0.0762$

33. 39.5 N · m

35. **a.** $60 \times 10^{+4} \, \text{m}^2$ **b.** $5.76 \times 10^3 \, \text{in.}^2$ **c.** $1.43 \times 10^{-3} \, \text{mi}^2$
   d. 106 L   **e.** $4.66 \times 10^3 \, \text{cm}^3$ **f.** $78 \times 10^{-3} \, \text{mL}$

37. 12.5%, 47.4%

39. 651 mL

41. $1.90 per gallon

43. #2/0 = 8.41 mm, #1/0 = 7.77 mm, #1 = 0.283 in.
#2 = 0.262 in., #18 = 1.22 mm, #18 1/2 = 1.12 mm
#19 = 0.0409 in., #19 1/2 = 0.0382 in., #20 = 0.889 mm

45. All dimensions are in mm. Accuracies vary by the number of significant figures in the inch measurements.

NEMA Frame	AH	AJ	AK	BB Min.	BD Max.
42	33.3	95.3	76	4.1	127
48	42.9	95.3	76	4.1	143
56	52.3	149	110	4.1	165

47. $\dfrac{23}{32}$ in. $\approx$ 18 mm $\pm$ 0.8 mm $\left(\dfrac{1}{32}\text{ in.}\right)$

49. For doors 32″ × 80″, the appropriate metric conversion is 0.813 × 2.00 meters $\pm$ 3 mm $\left(\dfrac{1}{8}\text{ inch}\right)$.

Chapter 6

1. a. $c = 10$ **b.** $c = 0.0985$ **c.** $a = 21.3$ **d.** $b = 0.968$ **e.** $b = 5$ mi

3. a. 25°   **b.** 39.1°   **c.** 82.85°   **d.** 85.46°   **e.** 89.9955°

5. sin 45° = cos 45° because both legs are equal
tan 45° = a/b and, again, both legs are equal

7. a. sin = 0.208, cos = 0.978, tan = 0.213
b. sin = 0.602, cos = 0.799, tan = 0.754
c. sin = 0.999, cos = 0.0349, tan = 28.6
d. sin = 2.09×10^{-4}, cos = 1.0000, tan = 2.09×10^{-4} (the sine and tangent are equal for small angles)

9. a. $\sin^{-1} = 30°$, $\cos^{-1} = 60°$, $\tan^{-1} = 26.6°$
b. $\sin^{-1} = 90°$, $\cos^{-1} = 0°$, $\tan^{-1} = 45°$
c. error   error $\tan^{-1} = 54.1°$
d. $\sin^{-1} = 60°$, $\cos^{-1} = 30°$, $\tan^{-1} = 40.9°$
e. $\sin^{-1} = 45°$, $\cos^{-1} = 45°$, $\tan^{-1} = 35.3°$
f. $\sin^{-1} = 87.4°$, $\cos^{-1} = 2.56°$, $\tan^{-1} = 45°$

11. a. $A = 36.9°$ (use $\sin^{-1}$)   **b.** $B = 65.2°$ **c.** $c = 22.5$ **d.** $b = c \times \sin B = 17.7$
e. $B = 36.1°$ (complementary rule), $a = b \times \tan A = 0.0041$

13. Refer to end of Appendix C.

15. tan 8° = 0.140; the tangent function's definition is equivalent to the definition of percent grade (percent grade = 0.140 × 100 = 14.0%)

17. Length of each wire = 55.3 m, angle of wires to ground = 40.6°, angle of wires to tower is 49.4°.

19. $\tan a = \dfrac{\text{opp}}{\text{adj}} = \dfrac{a}{\text{portion of base}}$, and $\tan^{-1} \dfrac{a}{\text{base portion}} = $ angle $A = 24.2°$

21. (2791,1100)

23. a. $X = 253, Y = 253$ **b.** $X = 21.4, Y = 12.4$ **c.** $X = 0.241, Y = 6.90$
     d. $X = 0.354, Y = 0.001$ **e.** $X = 3490, Y = 0$ **f.** $X = 0.110, Y = 0.0587$

25. 3.78×10^4 N

27. $\theta = \dfrac{7 \times 10^{-3}\text{s}}{1} \times \dfrac{1 \; \cancel{\text{cycle}}}{16.7 \times 10^{-3}\text{s}} \times \dfrac{360°}{\cancel{\text{cycle}}} = 151°$

29. Experimental solution required.

31.

$A_{\text{square}} - A_{\text{circle}} = A_{\text{fillets}}$

$s^2 - \dfrac{\pi}{4}s^2 = A_{\text{fillets}}$ $(A_{\text{circle}} = (\pi/4)d^2)$

$s^2 \left[1 - (\pi/4)\right] = A_{\text{fillets}}$, and $[1 - (\pi/4)] = 0.2146$

Therefore, $0.215 \, s^2 = $ area of four fillets.

33. Requires drawing to scale.

35. 48 ft^2

37. DC voltmeter reads $V_{\text{average}} \cdot V_{\text{av}} = 0.5 \, V_{\text{peak}} = 0.5 \times 5 = 2.5 \, V$

39. Use $V = \pi r^2 h$ and $A = 2\pi r(r + h)$
   a. $V = 0.0380 \text{ in.}^3, A = 0.767 \text{ in.}^2$ (shell casing capped at open end)
   b. $V = 25\,100 \text{ ft}^3, A = 5027 \text{ ft}^2$
   c. $V = 17\,700 \text{ m}^3, A = 3770 \text{ m}^2$

41. 7238 ft^3

43. $A_{\text{flange}} = 14.4 \text{ in.}^2$, area of two flanges $= 28.8 \text{ in.}^2$
   $A_{\text{web}} = 12.7 \text{ in.}^2$, total cross-sectional area of beam $= 41.5 \text{ in.}^2$

45. Requires individual response.

47. Requires individual response.

Chapter 7

1. See "Introduction," Chapter 7. Requires individual response.

3. a. G, S   **b.** G   **c.** G, S (similar to not zeroing meter)   **d.** G, S
   e. R   **f.** R   **g.** S, R (systematic only in tightly controlled experiments)
   h. S   **i.** S   **j.** R   **k.** G, S

5. a. 2   **b.** 2   **c.** 5   **d.** 5   **e.** 1   **f.** 2   **g.** 4
   h. 4 (decimal point implies accuracy to unit place)
   i. 5   **j.** 5   **k.** 6   **l.** 3   **m.** 5   **n.** 6   **o.** 5

7. a. 2.7×10^3 **b.** 4.4×10^3 **c.** $4.900\,6 \times 10^2$ **d.** 8.7000×10^3
   e. 6×10^4 **f.** 7.1×10^{-2} **g.** 7.800×10^{-1} **h.** 7.000×10^3
   i. 2.0058×10^2 **j.** 6.5893×10^2 **k.** $6.000\,07 \times 10^4$ **l.** 9.05×10^{-2}
   m. 3.0000×10^3 **n.** 3.0056×10^0 **o.** 2.7900×10^{-3}

9. $84.5 \pm 5 = 79.5$ to 89.5

11. a. 993 to 1007   **b.** 498 to 502   **c.** 39.1 to 40.9   **d.** 31.90 to 32.14 mA

13. a. 60   **b.** 990×10^3 **c.** 0.1   **d.** 9.27×10^{-3} **e.** 100
   f. 700×10^{-6} (engineering notation)

15. a. 3330   (the   0   is   not   significant,   and   no   decimal   point   follows)
   b. 3999.7   **c.** 4.0   **d.** 7362

17. 34.43 MPa

19. a. Point 1—graphing check list
   b. Point 5   **c.** Points 2 and 4
   d. Points 2 (*Y*-axis should be broken below 100) and 3 (flow is *dependent* on pressure and should be placed on the *Y*-axis)

21. 1.8 GPM

23. 0.9 A

25. Experimental report required.

Chapter 8

1. Mainframes (largest—data processing), minis (medium size—data processing and control), micros (single chip processor—low-cost data-processing and interfacing device [e.g., information from programmable logic controllers is compiled in the PC and sent on to the minicomputer]).

3. The minicomputer has three to four microprocessors to the PCs one. Minicomputers also have expanded memory and more powerful (faster) compilers. The PC is more portable and quite adequate for many industrial tasks.

5. c

7. d

9. d

11. Research and group interaction required.

13. d, E

15. b, C

17. e, A

19. Hardware = all fabricated computer parts; software = all computer information (e.g., programs, data input, etc.).

21. Central processing unit.

23. Requires individual response.

25. $\dfrac{54 \text{ characters}}{1 \text{ second}} \times \dfrac{1 \text{ line}}{60 \text{ characters}} \times \dfrac{1 \text{ page}}{25 \text{ lines}} = 0.036 \dfrac{\text{pages}}{\text{second}}$

   Time required would be 139 s or a little over 2 min.

27. Requires application software research.

29. Item 5, type material to purchase (#3 reinforcing rod), unit price based on foot of length, average unit price = $0.17, price variation between companies = $0.03, minimum price = \$0.12/ft, maximum price = \$0.22/ft; nine companies were surveyed in the nine-county area.

31. See Section 8.3. May require some outside research or classroom discussion.

33. Requires individual response.

35. Requires internet research.

37. Anything below 1024×768 pixels is considered low resolution.

39. See Section 8.5

41. Requires a visit to a local PC dealership.

43. No report required, but class discussion would be helpful here.

Chapter 9

1. Syntax error is an input the computer does not understand. The operator must check spelling and word order, then try a new input.

3. $2^{32} = 4.3 \times 10^9$, or 4.3 billion bits

5. See Section 9.1, and check your computer operator's manual.

7. ? $2 + 6 / 3 - 2$ [RETURN/ENTER] 2, with calculator: 2 ⊞ 6 ⊡ 3 ⊟ 2 ⊜ 2

9. ? $15.6E3*12E6*32E1$ [RETURN/ENTER] 5.99E13, with calculator: 15.6 ⌈EE⌉ 3 ⌈×⌉ 12 ⌈EE⌉ 6 ⌈×⌉ 32 ⌈EE⌉ 1 ⌈=⌉ 5.99×10^{13}

11. ? SQR $(32^\wedge 2 + 18 ^\wedge 2)$ [RETURN/ENTER] 36.7, or
? $(32 ^\wedge 2 = 18 ^\wedge 2) ^\wedge (1/2)$ [RETURN/ENTER] 36.7,
with calculator: 32 ⌈x²⌉ ⊞ 18 ⌈x²⌉ ⊜ ⌈√x⌉ ⊜ 36.7

13. D$ = $3 \times 4 \times 5 = 60$

15. "YOUR NAME" scrolls continuously down the screen. The result of a loop is to continually return control to a previous program line; thus, the program never stops running.

17. $T = 825$ in.·lb

19. 80 $B = 65000 * (3.1415926/4 * D ^\wedge 2)$
90 PRINT
100 PRINT "the breaking strength in lbs is"; B
110 IF B => T THEN 130
120 END
130 PRINT "The bolt will break"
The program shows no bolts will break.

21. COS (.524) [RETURN/ENTER]
= 0.866   (0.524 rad = 30°)

23. ASIN (.5) [RETURN/ENTER] 0.524 rad * 180/3.14159 [RETURN/ENTER] = 30°

25. 0.785 rad or 45°

27. $5\angle36.9°$; 3.54, 3.54

29. Requires computer language research.

31. The "set point" is a point in the process-control loop, usually an entry device such as a keyboard, where desired sensor values may be set. An example is your home thermostat with a sliding scale pointer.

33. Requires library research.

35. Refer to this chapter, classroom discussion, and your college library.

37. Moving parts are less reliable, require more power to operate, and are difficult to connect (and modify once connected).

39. The motor will run until the start pushbutton is pressed.

41. Requires in-class discussion.

43. Digital voltmeter, or DVM. Plotters, printers, and disk drives are examples of mechanical devices making up a complete computer system. CpETs must be able to repair both electronic and mechanical systems.

Chapter 10

1. Requires personal research.

3. Cybernetics is the science dedicated to discovering the common elements of the human nervous system. Cybernetics research will be used to build improved AI machines. Heuristic programming uses the "rules of thumb" of human experience—based on peculiar human traits such as judgment, intuition, and common sense—to write expert programs.

5. Requires personal research.

7. Requires individual response.

9. End effector, interfaces with work; arm, transports and positions end effector; controller, controls end effector and arm movements; power supply, furnishes electrical and/or fluid power to the robotic system.

11. Requires personal design work.

13. \$100/W = 0.01 W/\$ and \$6/W = 0.0167 W/\$. Therefore, $\frac{0.0167-0.01}{0.01} \times 100 = 83\%$ decrease in cost.

15. Requires library research.

17. Requires individual response.

19. Ceramics are tough, durable, corrosion-proof, light, and good heat conductors. Typical uses will be electronic packaging (especially heat-dissipating substrate bases), machine tooling, automobile engine parts, and high-temperature superconductivity.

21. To know how technology can improve the environment; and to protect the environment from the misuse of technology.

23. Requires individual response.

25. Requires library research.

27. Requires tour and individual response.

Index